AIR DISASTERS

Previous page:
Prototype Comet 1 G-ALVG — note Speedbird emblem.   *BAe*

Above:
Japan Air Lines 747.   *Via John Stroud*

AIR DISASTERS

STANLEY STEWART

LONDON

IAN ALLAN LTD

First published 1986

ISBN 0 7110 1585 6

Published by Ian Allan Ltd, Shepperton, Surrey;
and printed by Ian Allan Printing Ltd at their works
at Coombelands in Runnymede, England

Contents

Acknowledgements

A great deal of research and effort has gone into the production of this book and the author is deeply indebted to many people for their willing and enthusiastic help. All those who so freely contributed of their time and knowledge have been listed below, and to each and every one I would like to give my heartfelt thanks. A few individuals helped more than I deserved and it would be an injustice to their efforts not to make a special mention of their contributions. I must thank very much John Stroud, air transport writer and consultant, editor and author of many aviation books, for his enormous help and encouragement. His conscientious scrutiny of the material and his knowledge and advice on a great many matters has been invaluable. Without his efforts this book would be a sad reflection of its present form. I must also thank very much Sir Peter Masefield, industrialist, aviation expert and author. Over the decades the work and effort of Sir Peter in striving to establish the cause of the R101 disaster must be admired. His book on the subject *To Ride the Storm: the Story of the Airship R101*, and his *Aeronautical Journal* article, are the culmination of over 30 years of research. To both these gentlemen I must say a very special thank you.

I must also thank Andrea Thomas for her stalwart reading of the typescript. Her contribution to the final copy is immeasurable. Thanks also go to Sara Phillips for her careful and conscientious work in transposing the manuscript to the word processor. To both these ladies I must also say a special thank you. Many thanks are also due to:

R101 Geoffrey Chamberlain (*Airships: Cardington*); photograph on page 21 courtesy Public Records Office, Kew; **Comets** John Sturgeon (retired Farnborough scientist), Gordon Haydon (ex-Comet flight engineer) and Monty Montague and Doug Williamson (BA flight crew lecturers); **Munich** Mrs Ruby Thain (widow of Captain James Thain); **Trident (Staines)** John Sturgeon, Captain Deryk Ford (former Trident pilot), First Officer John Scott (former Trident pilot), Chris Howells (experienced light aircraft pilot); **DC-10 (Paris)** Flight Engineer Tony Haig (BA project on Paris accident); **DC-10 (Erebus)** Captain Gordon Vette (*Impact Erebus*); **Korean 747** Captain Mike Chan-Choong (former 747 pilot)

To all those mentioned I must express my thanks. Any errors remaining in the book are, of course, entirely my own.

Stanley Stewart

Abbreviations and Glossary

AINS	Area Inertial Navigation System
ASR	Area Surveillance Radar
ATC	Air Traffic Control
BEA	British European Airways
CVR	Cockpit Voice Recorder
FDR	Flight Data Recorder
F/E	Flight Engineer
FIR	Flight Information Region
F/O	First Officer
GMT	Greenwich Mean Time
HF	High Frequency
IFR	Instrument Flight Rules
IMC	Instrument Meteorological Conditions
INS	Inertial Navigation System
KLM	Koniges Lucht Macht — Royal Dutch Airlines
kt	knots
local	Local time if different to GMT
MHz	Megahertz
NDB	Non-Directional Beacon
nm	nautical mile
Phonetic Alphabet	Alphabet used for clear and distinct spelling. This is also used as a shorthand when referring to aircraft, so G-ARPI becomes 'Papa India'
QNH	Mean Sea Level Altimeter Pressure Setting
R/T	Radio Telephony
rpm	revolutions per minute
S/O	Second Officer
Tel	Telephone
Time	Reported in this book by means of the 24-hour clock. So 16.16hrs is 4.16pm and 16.16:23 is 23 seconds past 4.16pm
V1	Decision speed – the go or no-go decision point on take-off
V2	Minimum safe speed required in the air after an engine failure at V1
VFR	Visual Flight Rules
VHF	Very High Frequency
VMC	Visual Meteorological Conditions
VOR	VHF Omnidirectional range Radar beacon

Introduction

The events described in this book, although concerned with past tragedies, are also about hopes for the future, and attempts have been made to reveal the improvements to aviation brought about by such incidents as well as the causes of the accidents themselves. While dealing essentially with the disastrous loss of machines, these accounts are also about people, and often very interesting people at that. Although most of those mentioned perished in the accidents, their standing is not diminished by the telling of these tales. Many were very brave in the face of great danger. If lessons can be learned from previous catastrophes and can in any small way contribute to the prevention of their recurrence, then those who died so tragically will not have lost their lives in vain. Anomalous though it may seem, this book of accidents is also a book about safety.

Aviation accident fatalities for 1985 were the worst since records began, and one radio broadcaster was prompted to entitle a programme on the subject 'A Safer Way to Die'. In spite of this kind of cynicism, however, these numbers, although unacceptably high, can be compared favourably with other forms of transport. Whatever these accounts of disaster in this book may proclaim, flying, by any comparison, is still a remarkably risk-free form of travel. No excuse is offered for repeating statistics, at the beginning of a book on accidents, which declare flying to be so safe. Comparative studies of the travelling experiences of the average individual have shown that travel by train is the safest, followed by bus and aircraft. When comparing air travel to private motoring, however, the results are startling, even allowing for the fact that on average a person spends a lot more time driving than flying. In the United Kingdom alone over 6,000 people die on the roads each year, and in the United States the figure is over 30,000. The annual worldwide road traffic death toll is nearer 250,000, over 100 times greater than the worst aviation fatality total. If one is driven to an airport and picked up at the destination, the flight is the safest part of the trip: the odds of being killed between take-off and landing being calculated at three to one million.

On a 'distance travelled in a year' basis (statistics courtesy of Sir Peter Masefield and the Royal Society for the Prevention of Accidents), the risks of flying can also be shown favourably. If each one of a group of 100 people travels 10,000 miles each year, it can be calculated in years how long just one of that group of 100 people travelling by different forms of transport could journey before expecting to be killed. For scheduled airlines the figure is an amazing 940 years. The respective figures for other forms of transport are: buses and coaches, 600 years; British Rail, 550 years; charter airlines, 220 years; commercial vehicles, 150 years; private vehicles, 95 years; private aircraft, 20 years; pedal cycle, 10 years; and motorcycle, 5 years. These figures are somewhat unrealistic as one would not expect to travel 10,000 miles in a year by pedal cycle, but the details do reveal the safety of journey by air when viewed in the context for which the transport was intended: travelling quickly over long distances.

An examination of aviation fatalities against deaths from other causes demonstrates quite clearly the problems which exist in other fields. In the UK alone 100,000 people die from smoking-related diseases each year, a more virulent epidemic than the black plague in the Middle Ages, and the number of deaths from heart attacks and strokes each year (140,000) is equivalent to a packed 747 crashing every day. Although these statistics can offer little comfort to the relatives of those who have died in aircraft accidents, the great concern expressed in aviation at the comparatively small number of annual fatalities is a credit to the industry, and demonstrates quite clearly the true depth of feeling regarding aviation safety.

On 17 September 1908, Lt Thomas C. Selfridge was killed when Orville Wright's aircraft crashed after a wooden propeller splintered. Wright escaped with a broken leg. The unfortunate lieutenant held the distinction of being the first aeroplane passenger fatality. News of the accident was received with horror by the public and raised the whole question of safety for passenger-carrying aircraft. In World War 1, as aircraft came into wider use, attempts were made to reduce the accident rate. Reports of incidents were circulated amongst airmen to help improve knowledge. One RFC summary admonished that 'accidents during the last three months of 1917 cost £317 10s 6d — money down the drain, and sufficient to buy new gaiters and spurs for each and every pilot and observer in the service!' A statement on flight safety pronounced at the time clearly outlined the problem. 'Aviation in itself is not inherently dangerous. But to an even greater degree than the sea, it is terribly unforgiving of any carelessness, incapacity or neglect.' The same comment is as valid today as it was then, and over the years much effort has been expended in overcoming the problems.

Questions of safety are still being raised, and there is no doubt that the skies are becoming safer in spite of enormous and continuous increases in air travel. Thirty years ago the chance of a flight fatality was almost 10 times what it is now. Whatever is said about the hard-bitten fliers of yesteryear, fighting their way through the elements in bygone machines, the accident rate then was appalling compared to present standards. The modern airline pilot may be cosseted by electronics, but his or her contribution to safety is admirable. From July 1982 to May 1984, for example, US airlines carried more than 500 million passengers on eight million flights without a single fatality. Figures for the five years from 1980 to 1984 show that the average worldwide annual fatality rate was 940, with 1984 being the lowest since records began with only 451 deaths. Admittedly 1985 followed as the worst year on record with 2,129 killed, more than double the average, but these fatalities were the result of 40 accidents worldwide (in only two 747 losses — Air-India and Japan Air Lines — 849 people died), compared with 44 accidents in 1980 in which 1,329 were killed. Which was the safer year? In whatever light 1985 is viewed, however, the figures are disturbing, but travellers can be assured that lessons learned from these accidents will be vigorously applied and that in spite of a bad year safety will continue to improve.

All aviation personnel are normally carefully chosen and well trained in their various professions. All aircraft are normally serviced and maintained to the highest standards. Each individual is instilled with a high level of safety consciousness and everywhere attempts are made carefully to apply safety standards. The modern aviation world, however, is a complicated environment. Human beings, on occasions, make mistakes, equipment sometimes malfunctions and the fate of unfortunate timing is always ready to play a part. With the occurrence of a rare accident, news of the disaster is spilled upon a horrified world. The more people involved the more newsworthy the item, with the tragedy seeming to belie the statistics which proclaim flying to be so safe.

Accidents attributable to a single cause have been all but eliminated and air crashes now are more likely to be the result of a complex chain of events, tumbling one after the other, which lead to the sort of tragedies which are outlined in the pages of this book. In the day-to-day running of an airline, vigilance has to be the order of the day for every employee involved in front line operation, whether ground engineer, aircraft dispatcher, flight attendant or captain. If the recounting of these incidents can in any way improve attitudes to safety in the aviation community at large then the writing of this book will have been worthwhile. Rudyard Kipling, in his poem 'The Secret of Machines', wrote in a much more eloquent manner of the importance of vigilance and the risks of inattention. Here, the machines talk back to their operators:

We can pull and haul and push and lift and drive,
We can print and plough and weave and heat and light,
We can run and race and swim and fly and drive,
We can see and hear and count and read and write . . .

But remember, please, the Law by which we live,
We are not built to comprehend a lie.
We can neither love nor pity nor forgive —
If you make a slip in handling us, you die.

Ten Worst Accidents
1 Tenerife: KLM and Pan Am 747s collided, 1977. 582 killed.
2 Japan Air Lines 747 crashed near Tokyo, 1985. 520 killed.
3 Paris: Turkish DC-10 crash, 1974. 346 killed.
4 Air-India 747 off Irish Coast, 1985. 329 killed.
5 Saudi TriStar fire, Riyadh, 1980. 301 killed.
6 Chicago: American Airlines DC-10 crash, 1979. 271 killed.
7 Korean 747 shoot-down, 1983. 269 killed.
8 Air New Zealand DC-10, Mount Erebus, 1979. 257 killed.
9 Air-India 747 off Bombay coast, 1978. 213 killed.
10 Chartered DC-8 at Colombo, Sri Lanka, 1974. 191 killed.

The R101 Disaster   1930

In the quiet country churchyard of St Mary, Cardington, near the town of Bedford, England, lie together in a single grave the 48 airshipmen of R101. Heroes of a great empire, victims of a failed dream, these brave men were lost to the world, and to a noble cause, in the autumn of 1930. Also lost to Britain that year was an ambitious airship programme which was poised to revolutionise air travel, and was planned to bring to long-distance journeys a comfort and speed quite unknown in the past.

At the Royal Airship Works, within sight of the airshipmen's last resting place, stand to this day the two vast elongated hangars which dominated the airship scene. Lying side by side like giant cathedrals, the sheds mark the site of a story which began to unfold some eight years before the disaster. The following account tells of a wonderful dream to unite the far flung corners of the Empire by airship, of the enthusiasm for the cause, and of the courage and determination to see such hopes fulfilled.

A leading figure in the drama was the Secretary of State for Air, Christopher Thomson, the inspiration behind an ambitious airship programme which began in 1922. Created Lord Thomson by his friend Ramsay MacDonald, Prime Minister of the first Socialist government newly risen to power in 1924, the equally new lord had taken the title of Cardington to demonstrate his total commitment to airship development. The future of air transport seemed to be in the design of airships because, unlike aircraft of the day, they were peculiarly well adapted to long continuous flights over large distances, a useful advantage to the British Empire with its territories separated by great distances. The experimental programme adopted in 1924 called for the construction of two airships: the private enterprise R100, destined for Canada, to be built at Howden by The Airship Guarantee Co (a subsidiary of Vickers Ltd) and the Air Ministry R101, destined for India, to be built at the Royal Airship Works, Cardington. The two projects were arranged to be funded separately by the government as a means of ensuring competition in design and to encourage a healthy industry in both the public and private sectors. Such an arrangement also provided protection against one ship being lost in an accidental failure. Elaborate researches and experiments were to be made, including investigations into the new science of meteorology on the route to the Indian sub-continent, and sheds and masts were to be erected in England, Egypt, India and Canada.

The airship programme, planned with a number of other major objectives, formed the basis of an enthusiastic aviation policy which would keep Britain at the forefront of world aviation events. The proposed development of the airship, however, even by today's standards, was ambitious indeed, and lay at the heart of Lord Thomson's most eager hopes for the future. The requirement was for an airship of 30 tons payload (100 passengers and baggage, 14 tons; cargo, 16 tons), capable of flying non-stop, day and night, over distances of up to 3,500 miles, cruising at a speed of 63mph, in all weather experienced throughout the extent of the Empire. The necessary airship gas capacity would be of five million cubic feet, twice as large as any previously constructed, giving a gross lifting weight of about 150 tons. The resulting vessel was massive, even by comparison with modern big jets. When finally completed the total capacity of R101 (R stands for the rigid structure of the airship as opposed to a simple flexible construction) was five and a half million cubic feet. One million gold-beater's skins (the outer coat of the large intestine of the ox) were used for lining the gas bags. The outer cover stretched over five acres, the total modified length was 777ft 2½in, about three and a half times longer than the Boeing 747 jet (225ft 2in). The height was 139ft 1in, over twice the height (from wheel base to tail fin top — 63ft 5in) of the 747, and the width (131ft 9in), was about two thirds of the wing span (from wingtip to wingtip — 195ft 8in) of the 747. All who witnessed such a mammoth in the air, flying at heights of around 1,000ft above the ground, told of the experience with awe.

The lifting gas to be used in the British designs was hydrogen, as helium was not yet a commercially viable product in any quantity. Helium was produced naturally from oil wells in the United States and was employed as the lifting agent in some airships in that country, but the Americans were not able at that time, because of lack of supply, to make the gas readily available to everyone who required it. To improve safety on the British ships, a factor which had been given overriding consideration from the start, a new type of heavy-oil engine, the diesel engine, was to be developed for aviation use. Diesel oil was also safer to carry and was useful in hot climates where petrol would inevitably give off dangerous fumes. Diesel oil was cheaper, too, and fuel consumption less, allowing a reduction in weight with lower fuel requirement.

The first socialist government in Britain was short lived and fell in November 1924. Lord Christopher Thomson, however, in less than a year in office, had laid the foundations of a strong aviation policy that was to survive his demise and lead the way to the country's future in the air. It was five long years before Lord Thomson again took up his original appointment as Secretary of State for Air when the Labour government found itself re-elected in 1929. Construction of the R100 and R101 had proceeded in his absence from office but setbacks had caused delays and the programme was running more than three years behind schedule. Faith in the future of airships had not diminished, however, and confidence in the future had been expressed by Britain, the US and Germany, with the continuation of airship construction, although each country had also suffered disappointments. In comparison with the fledgling aeroplane industry airships were faring well. During the 11 years from 1919 until 1930 all seven attempted crossings of the Atlantic had been completed successfully without a single loss of life, an excellent record as compared with flights by aeroplane in which 27 attempts to cross the North Atlantic had resulted in 16 failures and 21 deaths.

The airship programme drawn up in 1924 had planned for the return flight to India to be completed by early 1927, but delays had resulted in the proposed

Indian flight being nearer 1930. On his return to office in the summer of 1929 Lord Thomson, therefore, had not only avoided missing the boat, or should we say airship, but had been reassigned his old post on the brink of all his hopes and aspirations for airship development being achieved.

Both airships were nearing completion: R100 at Howden, in Yorkshire, under the design of Barnes Wallis, assisted by Nevil Shute Norway, later to achieve success in the literary world as Nevil Shute; and R101 at Cardington, Bedfordshire, under the design of the Assistant Director/Technical (AD/T), Lt-Col Vincent Richmond, assisted by Sqn Ldr Michael Rope, widely considered to be a designer of genius. In overall command of both the R100 and R101 airship projects was the Director of Airship Development (DAD) Wg Cdr Reginald Colmore, a dedicated airshipman of proven technical and administrative ability. On the flying side, the officer in charge of flying and training was the Assistant Director/Air (AD/A) Maj George Scott, a veteran of non-rigid and rigid airship experience. Scott's first command was on the non-rigid British Naval Airship No 4 in 1915, followed later by command of HM Airship No 9 in 1917, the first British rigid airship to take to the air. Scott's crowning glory was in command of R34 which completed the first return flight across the North Atlantic in 1919, a historic achievement. By 1929, however, Maj Scott was not the man of his earlier years, and, perhaps unwell, had lost much of his reputation gained on the R34. A series of mishaps in command had tarnished his image and left his judgement in doubt amongst many of his colleagues. His contribution to airship development and his reputation as the most experienced British airship commander, however, stood him in good stead and secured his position at Cardington. Under Maj Scott's command were the two airship captains, Sqn Ldr Ralph Booth of the R100, known to his friends as 'Mouldy', and Flt Lt H. Carmichael Irwin of the R101, a former Olympic athlete and nicknamed 'Bird' from the way in which he ran. Both airshipmen were held in high regard, with 'Bird' Irwin being considered by many as one of the best airship captains.

By the autumn of 1929 it was becoming clear that R101 had a weight problem! The experimental Beardmore Tornado heavy-oil engines had been delivered producing less horsepower than had been anticipated and weighing more than double the original estimate. Although a lower fuel load could be carried with the reduced fuel consumption of the diesel engines there was still a net increase with the heavier machines. Calculations of the all-up weight of R101 gave an estimate of 113.4 tons, 23.4 tons over the original 90 tons specification, not only because of the increased weight of the engines, but owing also to the extra strong airframe structure and the generous passenger accommodation. The actual lifting capacity in standard atmospheric conditions was 146.3 tons, leaving a useful lift of only about 33 tons, almost half the original specification. Fuel requirements for long-range flights were of the order of 30 tons: water ballast, stores, crew and spares 20 tons: plus the weight of about 100 passengers and their baggage another 14 tons. Although R101, in all respects, was an experimental vessel, commercial considerations were uppermost in the designers' minds and clearly major alterations were required. Even a demonstration flight to India in this state would be impossible, never mind one that was commercially viable: less weight and more lift were essential.

In the meantime tests on the R101 in her present condition were to continue. The gas bags were duly filled, and the airship floated clear of its trestles in the shed. Provisions were made for up to 16 tons of water ballast to maintain equilibrium, arranged in 28 containers throughout the airship, half of which was available for quick discharge in the event of an emergency. Fuel was also distributed uniformly and could be pumped by compressed-air from tank to tank to trim the airship for level flight. While in the shed, temperature variation between day and night required the pumping of water to and from the vessel to keep the ship just airborne.

The walking of R101 from the hangar was delayed by bad weather until the still conditions of 12 October 1929, when the airship was at last taken from its shed for the first time. A cheer rose from the assembled crowd as it was secured to its 200ft high mooring mast. Two days later R101 slipped from the mast at Cardington on its maiden flight and headed south at 1,500ft to make a triumphal pass over Buckingham Palace, Westminster, St Pauls and the City, viewed by Lord Thomson from atop the Air Ministry building. The airship returned to base in just under six hours: a most successful flight. On board were all the senior Royal Airship Works personnel and high ranking Air Ministry staff, a total complement of 48, including passengers and crew: a quite unnecessary risk, by today's standards, on an initial experimental flight; but considered quite normal at the time.

The airship programme created much interest amongst the public and press and large numbers of people were attracted to Bedford to view R101 riding at the mast, not unlike the attention focused on Concorde today. On 18 October a crowd assembled once more to witness the departure of R101 on its second trial flight, this time with Lord Thomson on board. A large press corps was also in attendance to record the event. Arguments for and against airship development were frequently vented in papers and magazines, sometimes quite strongly, and the press, although normally friendly, could on occasion be rather hostile. Airships were designed, out of economic necessity, to fly at no more than 1,500ft. In spite of earlier successes the dangers inherent in such a mode of transport were obvious to all, and the wisdom of constructing such a large vessel to fly, by necessity, so close to the ground was often questioned; although, of course, airships were essentially intended for flights over water. The difficulties encountered by both airships in the programme were well known to the public, and R101's weight problem sometimes ridiculed.

On the return of R101 to Cardington after its second flight a group of eager newsmen awaited the Secretary of State for Air at the bottom of the tower. After comments to the assembled press about the comfort and enjoyment of the flight Lord Thomson re-emphasised the nature of the tests.

'I wish to make it perfectly clear', he said, 'as I did when I introduced the programme in 1924, that it is all tentative and experimental. So long as I am in charge no pressure will be brought to bear on the technical staff, or on anyone else, to undertake any long-distance flights until they are ready and all is completely in order.'

Whether Lord Thomson fully lived up to his statement or, in fact, did bring undue pressure to bear on those responsible, is an argument which continues to this day. Already, however, he seemed to be setting deadlines in his own mind.

'Subject to this,' he added, 'I hope that perhaps it may be possible for me to travel to India in R101 during the Parliamentary recess at Christmas.'

Lord Thomson had, of course, been fully informed of the difficulties surrounding R101 and even to suggest the possibility of an Indian trip in a couple of months seemed less than wise. Had he failed to grasp the immensity of the problem, or had there in fact been a failure to inform him, perhaps unintentionally by these proud airshipmen, of the full details of the troubles facing R101? It may, of course, simply have been a ruse to pacify the press as the airship programme had cost a great deal of money up to this point with few tangible results.

'But,' Lord Thomson concluded, 'whether this is possible or not, the whole policy of the airship programme is "Safety First" and "Safety Second" as well.'

Forthcoming events were to tell a different tale.

In spite of the difficulties that lay ahead for R101 the crew were most impressed with the airship's flying qualities. Unfortunately, the same was not being said of the abilities of Maj Scott, the Assistant Director/Air. Doubts expressed earlier as to Scott's judgement were beginning to prove well founded, and the first outward signs of dissension between the major and Flt Lt Irwin and the other officers were beginning to show. Efficient docking of an airship to the mast was not one of Scott's better skills, and just how difficult the task could be was demonstrated on more than one occasion by the major. On arrival at Cardington from the return flight with Lord Thomson he took command from 'Bird' Irwin and proceeded to make a mess of the operation, taking 2hr 20min to complete the task. Maj Scott, sadly, was proving, as suspected, to be less than equal to the task. Experience alone, unfortunately, is neither proof of ability nor guarantee of performance. Irwin and his First Officer, Noel Atherstone, were clearly upset by Scott's display, and it was not to be the only time.

Test flights continued with the usual complement of dignitaries being carried at unnecessary risk. On at least one occasion the crew felt that the flight should not have been made under any circumstances, never mind with important passengers on board. During this period R101 required constant attention from the flying staff, both whilst in the air and when moored to the mast, and the strain of the work load was beginning to tell. Scott's interfering habits and bad judgement continued to be evident and led to more incidents which further increased tension amongst the officers.

On 11 November, with R101 riding at the mast, a storm struck of such severity that shipping on the coast was badly affected and buildings ashore damaged. At the height of the gale, with gusts of up to 89mph, the worst encountered by an airship, R101 rode quietly at the mast, rolling gently, but under control. Stability of the airship in gale conditions had been proved, but inspection following the storm revealed that the gas bags had moved while she rolled, causing damage. The bags had chafed against airframe protrusions resulting in a considerable number of gas leaks, and had she encountered such turbulence in cruise the result could have been disastrous. The experience boded ill for the future.

In all, seven test flights of R101 in her original condition were completed before she was returned to the hangar for modifications. Apart from the major problems of excess weight and greater lift requirement the flying trials had gone well, the airship had handled satisfactorily, and fuel consumption calculations were excellent. The Cardington design team now threw their combined efforts behind the task of reducing weight and improving lift. A list of 20 weight saving suggestions was drawn up, and the decision also taken to slacken the gas bag corsetting to increase volume and therefore improve available lift. There was still the problem of chafing on the frame joints, of course, and the assistant designer, Rope, had actually suggested that, in certain areas, some tightening of the wiring constraining the gas bags might be necessary to prevent rubbing, so the suggested expansion created obvious problems. However, the need for more lift was an overriding factor and chaffing, it was argued, could always be prevented by padding the frame joints.

Among the weight saving proposals were removal of two water ballast tanks, two lavatories, the reefing girders at the bow, the servo steering gear, and some galley equipment. It was also proposed to reduce the number of double-berth cabins from 28 to 16, thus effecting a total weight reduction of over three tons. On the lift side, further trim tests in the hangar demonstrated R101 to have a lifting capacity of 148.6 tons. Slackening of the gas bag corsetting to allow the bags to fill the airframe space more completely was estimated to improve lifting capacity by 3.2 tons, giving a gross lift of 151.8 tons. When all was said and done this gave a practical disposable lift of around 42 tons, still quite inadequate for the requirements. Even if R101 could reach the proposed destination Karachi (at the time in India), with a refuelling stop in Ismailia, Egypt, the return flight would only barely be possible because of lift loss in the very warm air, and even that avoiding the hottest months.

The solution to the problem lay in increasing the volumetric capacity of the airship by inserting an extra bay into the structure, and by reducing engine weight, thus raising the disposable lift to 55 tons. As far as the designers were concerned there were no engineering or aerodynamic restrictions to lengthening the airship, and the proposal was presented to Lord Thomson for approval. The programme would, of course, incur additional delays, and could not be expected to be completed until the early summer of 1930. The monsoon weather commencing in India at that time excluded the prospect of an experimental flight during the hot summer months, leaving a date of late September as the most probable for the flight. Such a departure time also allowed the return of R101 to Cardington before the onset of winter weather in England.

Disappointed, but not entirely surprised, Lord Thomson agreed to the proposals. Besides a return flight to India by the following September would not be inopportune. The Imperial Conference was to be held in the autumn of 1930 where experience gained from the R100 and R101 flights, particularly as regards regular commercial airship overseas services, would be discussed. Also, during the summer of 1929, Lord Thomson had been requested by his friend Ramsay MacDonald to consider the post, subject to the King's approval, of Viceroy of India, a position which would be vacant with the completion of the present Viceroy's tour of office early in 1931. An appearance before the Imperial Conference to announce his appointment as Viceroy, having just returned from a successful first flight to India, would be triumph indeed!

16

Meanwhile R100, at Howden, was nearing its maiden flight. At quite an early stage in the programme a planned conversion of the second-hand Rolls-Royce Condor engines from petrol to hydrogen/kerosene was abandoned as impractical and precluded the use of R100 in hot climates. It was proposed, therefore, to restrict flights of the rival airship to Egypt in winter, and Canada in summer. As a result the trial flight was altered to the latter destination, while the aim of its competitor, R101, remained India.

On 16 December R100 was walked from the shed, and, under the command of Maj Scott, with R100's captain, Sqn Ldr Booth, in the temporary position of first officer, she flew up and away from Howden to her new home in Cardington some 150 miles away. On this maiden voyage, 14 passengers were carried, including all the senior Howden staff! Problems occurred with the outer cover ripping and splitting in a number of places, which gave cause for concern. On the following day the second test flight from Cardington had to be cut short because repairs had to be made when a panel on the fin broke loose. The year thus ended with both airships sitting snugly side by side in their respective sheds at the Royal Airship Works undergoing modifications, R101 having completed seven test flights and R100 two.

R100's trials resumed in January after preliminary repairs to the outer cover with disturbing results. The cover tore once again in a number of places, the gas bags became wet and the electrical system was affected. There was no option but to return R100 to the hangar for further repairs and she remained there alongside R101 until nearly the end of April. In the humid conditions of No 1 hangar R101 fared little better. Its outer cover had been known to leak, notably during the severe storm at the beginning of November the previous year, and the material had begun to deteriorate badly requiring reinforcement with fabric bands. Work also continued on the slackening of the gas bag corsetting while at the same time engine modifications were undertaken. The design office staff were also busy with plans for the additional bay as well as for the proposed new generation of bigger and better airships: the R102 and R103.

R100 resumed test flying at the end of May, but a failure to the pointed tail section soon had her back in the hangar. Fortunately, at this time, a request was received from Canada to postpone the flight until after the Canadian election on 28 July as many MPs wished to travel to Montreal to witness R100's arrival. A breathing space had been offered and gratefully received.

On 23 June, after seven months inside, R101 was at last taken from the shed in preparation for a brief appearance at the RAF Display at Hendon at the special request of the Chief of Air Staff. The gas bag wiring had been slackened but the lift increase was found to be disappointing, while just over two tons weight had been saved. The extra bay, however, had still to be inserted, and design work was almost complete. The condition of the outer fabric was still causing concern and, while riding at the mast, a number of splits occurred which had to be rapidly repaired for the forthcoming air show. In all, three flights were completed in June, including a pass over the RAF Display. On each occasion handling of the airship seemed heavy, especially during the latter part of the journey. The ship kept losing height and had to be pulled back up again. By the end of the month R101 was back in the hangar where inspections revealed that, as suspected, the gas bags had rubbed against the nuts, bolts and taper pins of the airframe causing many leaks; and this in spite of 4,000 joints being padded during its time in the shed. Here was a problem that would undoubtedly incur further delay, not counting work on the new bay as yet to be installed. The flight to India, now only three months away, hardly seemed possible. Lord Thomson, in a speech to assembled press representatives before the display at Hendon once again stressed 'that the men responsible for this very great experiment with airships are not going to be rattled into doing anything which is not strictly in accordance with the principle of "Safety First"; not while I am where I sit. We are not going to allow that ship its start while there is the least little defect at all perceptible'.

The time was fast approaching, however, when the nation, and the Treasury who funded the project, would demand results. What disappointment would be felt by the Secretary of State at news of yet another set back.

The Aeronautical Inspection Department's own inspector in charge at the Royal Airship Works, Frank McWade, expressed, in a letter to the Air Ministry, his own deep concern at the situation. McWade, an engineer with extensive experience in rigid and non-rigid airship construction, stated quite bluntly that in his view R101 was not airworthy and that he was not prepared to issue further 'Permits to Fly'. The Air Ministry, with little or no knowledge of airship design, immediately consulted Wg Cdr Colmore, the Director of Airship Development, regarding McWade's remarks. Colmore, perhaps influenced by the desperate need for extra lift, contradicted McWade's opinion, stating that there was no objection to padding of the airframe, even though thousands of holes were still occurring despite the already extensive padding. When wet, the gas bags scuffed against the damp padding making the situation worse. In the end McWade was overruled and further padding approved. The problem of gas leaks caused by rubbing and chafing was not deemed by the Air Ministry of sufficient importance to 'bother' Lord Thomson and he was never informed of McWade's report.

The Member of the Air Council responsible for airships, Sir John Higgins, equally unaware of any gas bag problem, wrote to Lord Thomson stating that work on the new bay should be completed by 22 September, and then only one trial flight would be necessary before departure for India at the end of September. On 14 July a further statement by Sir John outlined a possible additional setback. It was deemed prudent to hold R101 in reserve for the Montreal flight in case R100 proved unreliable, which would unavoidably delay insertion of the new bay. Lord Thomson's reply was simple and to the point.

'So long as R101 is ready to go to India by the last week in September this further delay in getting her altered may pass. I must insist on the programme for the India flight being adhered to as I have made my plans accordingly.'

How was such a statement to be interpreted? Was this consistent with an earlier comment that 'no pressure will be brought to bear on the staff'? Perhaps Lord Thomson's eagerness for the flight to India was beginning to influence this otherwise considerate gentleman. Perhaps his dream of returning to India, the land of his birth, in the following year, arriving aboard his newly established airship service to fill the post of Viceroy, the highest office in the land, was proving too much to bear. Perhaps, as has been suggested, he was simply confirming his wish for the programme to proceed as planned, rather than

18

insisting on a definite departure date. Whatever the intention of the statement, the message was relayed post haste by Sir John to Wg Cdr Colmore at Cardington.

To some of the staff at the Royal Airship Works, Thomson was more like a god than the Secretary of State for Air. From the beginning he had been a paragon of enthusiasm, encouragement and inspiration. His wish, to some extent, was their command. Apart from Lord Thomson's hopes, however, there was also the question of weather. If the timing of R101's departure was not just right there was the prospect of postponement of the flight for several months until after the worst of the European winter. In whatever way history might interpret Lord Thomson's assertion, the pressure at Cardington was most definitely on.

In the early hours of the morning of 29 July 1930, R100 slipped from the mast at Cardington for the start of its journey to Montreal. Only one test flight had been completed before departure and the trial had been far from satisfactory. The outer cover was still showing to be letting in water and one of the gas bags was leaking excessively. Repairs were hastily conducted at the mast and the Air Ministry was informed of Colmore's affirmation that all was well for the Atlantic crossing. The crew, of course, were superbly confident of the airship's capabilities. Maj Scott was to be in charge of flying, but the small matter of his poor performance had been cleared up and his position on board was to be that of 'non-executive Admiral'. Sqn Ldr Booth was most certainly to be the captain. However, whatever the official outward show of confidence by Colmore, on the evening before departure he 'was in a very disturbed state'.

R100 arrived in Montreal in the early hours of 1 August after a not uneventful flight. Several tears in the outer cover had occurred requiring running repairs and the airship had been tossed violently in a line squall. On the evening of 13 August, after a number of days at the mast for repairs, followed by a local flight over Ottawa and Toronto during which an engine broke down beyond repair, R100 set off for the return flight to Cardington. At the beginning of the Atlantic crossing the airship ran into heavy rain which soaked the outer cover, wetting the crew's quarters and shorting the electrical and heating equipment. The temperature of the cabin and the food remained cold for the rest of the flight. At the end of the crossing, just south of Ireland, a young rigger by the name of Patterson was sent to relieve the rather tired height coxswain who had the unenviable and exhausting task of holding altitude for long periods without a break. About an hour later fatigue also affected Patterson and he dozed off, letting the ship plummet from 2,000ft to 500ft before Booth could grab the wheel and pull her out of the dive. A near disaster!

At 11.00hrs on Saturday 16 August R100 docked at the mast. The masthead platform was full of people waiting to meet the crew, among them the Secretary of State for Air, Lord Thomson, and the Director of Civil Aviation, Sir Sefton Brancker. R100 had returned in triumph, but not without problems. The outer covers and gas bags were in a terrible state, and much work was required on the engines. Early next day R100 was lowered from the masthead and placed in No 2 hangar, never to fly again.

After an extensive refit lasting three months R101 was at last handed over to the flying staff on Saturday 27 September 1930. The new bay had been inserted

Above:
R101 at the mooring mast. *Courtesy Geoffrey Chamberlain*

Below:
Control car command to nose riggers. *Courtesy Geoffrey Chamberlain*

Below right:
Control car engine telegraphs. *Courtesy Geoffrey Chamberlain*

Right:
Facsimile of the R101's Certificate of Airworthiness.

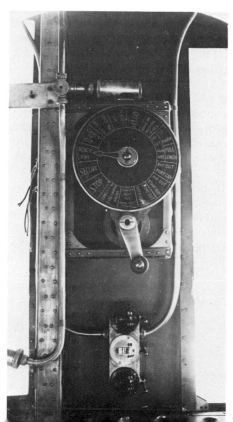

C.A. Form 82.

GREAT BRITAIN.

AIR MINISTRY.

CERTIFICATE OF AIRWORTHINESS No. L/A 6.

(Airships.)

FIRST PART.

FULL NAME, ADDRESS AND NATIONALITY OF OWNER OR OF OWNING COMPANY.

1. { Surname of owner (or name of Company) : **The Air Council.**

   { Christian name :

2. Address : **Whitehall, London, S.W.1.**

3. Nationality : **British**

NAME OF CONSTRUCTOR.

4. Name : **Air Ministry (Royal Airship Works)**

NATIONALITY AND REGISTRATION MARKS.

5. .... **G-FAAW**

DESCRIPTION OF AIRSHIP.

6. Type **R.101.** Constructor's No. **-**

7. Place of construction and year of completion of airship : **Cardington, 1929.**

8. Usual Station : **Royal Airship Works, Cardington, Bedford.**

9. Class of Airship **Rigid**
   (Non-rigid, semi-rigid or rigid)

10. Classification of Airship { Category : **Special**
    { Subdivision : **(v) Research or experimental**

11. Maximum Gas Capacity : **5,508,800 cubic feet.**

12. Overall length of airship : **777 ft 2½ inches.**

13. Overall height of airship (with and without landing shock absorber) : **141 ft. 7 inches 139 ft. 1 inch.**

14. Overall width of airship : **131 ft. 9 inches.**

21

increasing the length by 53ft from 724ft to 777ft. The gross lift had been increased to 167.2 tons, with the fixed weight now 117.9 tons, giving a 14.4 tons increase in disposable lift to 49.3 tons. A new outer cover had been fitted, except in certain areas of the nose and tail which had originally been doped after fitment and which were considered sound. Although not entirely satisfactory there simply wasn't time to renew the total cover area. Padding of the joints had been improved with the use of fabric bandages which even the discerning chief inspector at Cardington, Frank McWade, approved as satisfactory. Installation of two new reversible engines for use when docking had been completed and the gas bags had been repaired and strengthened. In spite of earlier setbacks the work had been finished with enthusiasm and confidence and R101 was now in better condition than either of the two ships had been at any time.

Now ready, R101 spent a number of frustrating days in the hangar awaiting calm conditions. Finally, on 1 October, the airship was attached to the mast and all preparations were made for the test flight. The handling party then walked R100 from No 2 shed to the larger No 1 shed in preparation for insertion of its new bay, the only time at which the two airships were in the open together. R101, now riding at the mast, was virtually a new ship. In such circumstances a whole series of tests of varying lengths in adverse weather conditions would normally be required, yet only one test flight was proposed before the 5,000-mile journey to India. Even as early as June, Sir John Higgins conveyed the impression to Lord Thomson that only one test flight would be required after refit, yet not one word from anyone at Cardington had contradicted the proposal and insisted on additional trials. Now it was even intended to reduce the original 24-hour trial flight to the minimum required in order to be able to return to Cardington, Air Vice-Marshal Hugh Dowding, the new member of the Air Council responsible for airships to succeed Sir John, in time for an important appointment. Dowding, in fact, had made it clear to Colmore that the duration of the trial was not to be reduced on his account, so there was no excuse for not flying the full programme.

On 1 October the Prime Minister, Ramsay MacDonald, opened the Imperial Conference with an announcement that R101 was at the masthead awaiting air tests prior to her journey to India. That afternoon R101 eventually took to the air in excellent weather conditions for its one and only test flight with a grand total of 57 aboard. Among the passengers was the usual sprinkling of dignitaries, including Air Vice-Marshal Dowding, the man now responsible for airships, who had never before been in an airship in his life! Sixteen hours later R101 returned to the mast at Cardington, the ship was pronounced to be fit and well, and was passed as in a satisfactory condition for the issuing of the Certificate of Airworthiness. There had been no bad weather or turbulence trials, no endurance tests, nor any high-speed checks because early on an air cooler had failed, which resulted in one engine being shut down. The flight was undertaken in near perfect conditions and amounted to little more than a joy ride, yet it was apparently considered sufficient to assess the airship as satisfactory. The year before Flt Lt Irwin himself had drawn up a suggested programme of checks for airship trials which had been disregarded. He, if anyone, must have had some reservations. However, on the surface at least, the atmosphere was one of optimism and confidence.

22

Departure for the voyage to India was set for 19.00hrs on the evening of 4 October and a request was made for Lord Thomson and his distinguished party to be in attendance at Cardington 45min earlier at 18.15hrs. Loading, fuelling, gassing and last minute repairs to R101 were conducted throughout the day and every effort was made to save each ounce of weight with all baggage being restricted and carefully weighed. Fuel on board amounted to 25 tons giving a still air range of 3,200nm. With engines set to long-range cruise a 2,420nm range was possible against an average 20kt head wind: and that distance, in what was reckoned to be the worst conditions expected, was still 185nm longer than the Cardington to Ismailia sector. The weather forecast at first promised good prospects but a further report showed some deterioration by evening. A depression west of Ireland was moving east bringing adverse southwesterly winds at 15mph with rain later but it was hoped to schedule departure before any break in conditions. Sqn Ldr Johnston, the chief navigator, had plotted a course to London, then over Kent to Hastings and on across the Channel. The route over France was planned to pass just north of Paris then on to Toulouse flying west of the Rhône valley, leaving the south coast at Narbonne. Turning east the airship would then be free of the forecast head winds with calm conditions and the safety of the flat Mediterranean lying ahead.

By 18.00hrs the last of the loading was being completed and the time to board all personnel was approaching. A huge crowd had gathered round the perimeter to witness R101's departure in the approaching dusk, some having camped out all night to secure a view. At 18.15hrs Lord Thomson was driven to the base of the tower followed closely by his luggage in a Ministry van! The baggage and accoutrements weighed a grand total of 254lb, including two cases of champagne at 52lb and a Persian carpet at 129lb. The carpet had been presented to the Secretary of State in Kurdistan in 1924 when he was on an official visit to RAF Command in Iraq and with the champagne, was intended to grace R101's lounge at the formal dinners planned for Ismailia and Karachi. The baggage was duly loaded but the carpet, being 10ft long, was placed for easy stowage along the gangway right up by the nose. At the base of the tower the remaining boarders, the designer Lt-Col Richmond, flanked by Lord Thomson and Sir Sefton Brancker the Director of Civil Aviation, stood for a final photograph. At the last moment before departure, unbelievable as it may seem, the Certificate of Airworthiness (C of A), without which the flight was not permitted, was finally handed over.

With the load sheet signed and the C of A in his possession, Flt Lt Irwin ascended the tower in the lift for the last time and climbed aboard. The total ship's complement amounted to 54 souls, comprising six passengers, six Royal Airship Works senior staff, five officers and 37 crew. As each of the five engines was started and roared into life the belly of the airship glowed in the floodlights from the mast. Finally, at 18.36hrs GMT, the restraining line was detached and the airship immediately dipped forward displaying her nose-heavy conditions. A fine mist of water fell from the front of the airship in the bright light as all forward ballast, except the two half-ton ballast bags jettisonable only at the nose, was released to gain lift, and, with a cheer from the crowd, R101 rose from Cardington and climbed slowly into the air. The airship made a traditional circuit

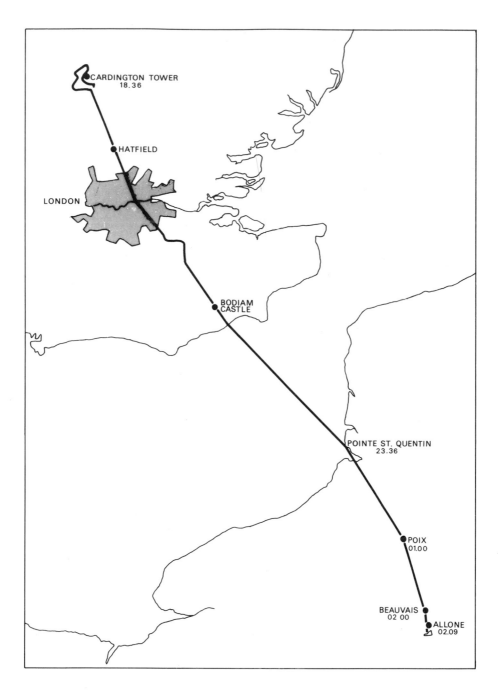

Above:
The track of the last flight of R101.

of Bedford and then, with rain beginning to fall and the wind strengthening from the southwest, set course on her way to Egypt and India.

Shortly after departure, within 30 minutes of slipping the mast, No 5 engine in the aft car displayed oil pressure problems and had to be shut down. As work proceeded on the failed machine the weather continued to worsen. Staying low to minimise gas loss and avoid the strongest winds, R101 droned on through the stormy night at an average ground speed of 39mph, flying at heights not much above 1,000ft. Thousands emerged from their homes to watch her passing. While the privileged few on board sat down to a cold supper, No 5 engine was restarted but shortly afterwards was shut down again due to the same problem. At 20.21hrs the following signal was transmitted prefixed with the four engines running code. 'Over London. All well. Moderate rain. Base of cloud 1,500ft. Wind 240° at 25mph. Course now set for Paris. Intended to proceed via Paris, Tours, Toulouse and Narbonne. Ends.' The airship continued over London and on in the pouring rain directly over the Royal Naval College at Greenwich.

Now two hours into the journey, with London behind and one engine failed, R101 was encountering severe weather. Strong, gusting winds, turbulence and heavy rain battered the ship, soaking the outer cover and adding more than three tons to the all-up weight. To make matters worse the latest weather forecast received from Cardington indicated deteriorating weather over France with rain, low cloud, and winds of up to 50mph. Caution might have demanded the return of the flight to Cardington, and had the Captain, Flt Lt Irwin, had his proper say that might well have been the outcome, although it would have been a brave decision to steer back through the storm to a somewhat ignominious return. Maj Scott, however, had his own ideas, driving the ship on at fast cruise in order to clear the weather as quickly as possible. Although by such a plan time within the storm area might be reduced, at high speed stresses on the aircraft and outer cover were increased considerably in the buffeting. While rolling, the sensitive side-mounted valves would undoubtedly be spilling some gas and the airship would be losing valuable lift.

R101 crossed the coast of England a few miles to the east of Hastings, witnessed from the cliff tops by a crowd braving the elements on the wind-swept night. Passengers could be seen framed in the observation windows, and to the watchers on the ground her passage seemed majestic as she crabbed through the air. The time was 21.35hrs, three hours, all but one minute, from the Cardington departure in which a distance of 127 miles had been covered at an average ground speed of just over 42mph.

Out over the Channel, now far from sight of pin-point landmarks on the ground, master navigator Johnston had to resort to 'dead reckoning' to calculate position. As an aid to navigation a racked box in the control car contained highly inflammable calcium flares which could be dropped into the sea at regular intervals. On contact with water the flares ignited to emit an orange plume which allowed drift calculations to be made. As flares were dropped on the crossing at least one watcher on the now distant cliff felt concerned at the orange glow and ran to telephone the lifeboat service.

The airship seemed to the crew to be coping with the adverse weather and the gas bags appeared to be holding up well in spite of the turbulent conditions, although the ship was taking a lot of punishment. Gas leakage had been previously estimated to result in a lift loss just a bit less than weight loss with fuel consumption, and so was not expected to pose a problem. In fact water ballast was actually being replenished from the rain using a unique water collecting system of channels on the top of the cover. Taking all factors into account, therefore, including the additional weight of the soaked outer cover, the water ballast recovered from the rain, and the loss of lift from gas seepage, R101 was estimated to be flying at about three tons heavy. In other words, the airship weight was greater than the gross lifting capacity of the gas bags by three tons, but was more than overcome by the aerodynamic lift from the hull at cruising speed. As long as power was maintained to hold speed, sufficient lift was available. In fact earlier stability calculations made by the designers had shown that even if a forward gas bag deflated sufficient control would be available provided engine power was sustained.

Fuel calculations, too, were encouraging. Even if the present conditions were to continue it was shown that sufficient fuel was available to complete the flight to Ismailia. Wind strength, however, was increasing and R101 was being blown steadily eastwards from its intended direct track to Paris. The airship was undergoing the severest of tests in worsening weather. At a request to the French a handling party had been positioned at the then military aerodrome of Orly for use in the case of an emergency, and if any untoward incident occurred before then they might just prove their worth.

Shortly before 22.00hrs, two meteorological messages received in quick succession from Cardington forecast tail winds from Paris to the south coast with fair weather beyond, and brought a measure of relief to the situation. By mid-Channel repairs to No 5 engine had been completed, and, approaching the end of the first night watch at 23.00hrs, the engine was started and the airship restored to five good engines. The way ahead now seemed assured and the decision to press on vindicated.

'Bird' Irwin assumed command from his First Officer, Atherstone, and settled into the middle watch. He, like all the members of the crew, was somewhat fatigued by the almost constant period of duty during preparation, departure and first phase of the flight, and looked forward to a well earned rest at the end of this three-hour period of duty. Radio bearings from Le Bourget and Valenciennes were obtained to indicate position, and then, at 23.36hrs, exactly five hours after departure, R101 crossed the coast of France at Pointe de St Quentain on the mouth of the Somme giving a precise position fix. The airship was seen to be 20 miles off-track and a heading of 200° was calculated by the navigator, Johnston, to make good the direct track to Paris. The intended flight path thus lay about four miles west of Beauvais, but an incorrect calculation of wind resulted in further easterly drift being experienced routeing the airship over the village of Allonne by the Beauvais Ridge which lay ahead. The area was not unfamiliar to one distinguished guest, Sir Sefton Brancker, the Director of Civil Aviation. In 1928 aboard an Argosy of Imperial Airways he had experienced a violent downdraught in the vicinity of Beauvais with winds from a similar direction. On that occasion passengers had struck the roof, seats were broken, and an exit door was torn off.

26

Just before midnight a further bearing from Le Bourget indicated the drift to be greater than expected but in the fatigued control car no heading correction was requested to allow for the oversight. At 24.00hrs the following message was transmitted by Irwin to Cardington: 'Position: 15 miles southwest of Abbeville. Course and speed made good from 18.30hrs: various headings and 33kt. Wind: 245°, 35mph. Altimeter height: 1,500ft. Air temperature: 51°F. Weather: intermittent rain. Cloud: Nimbus at 500ft. Conditions since departure: similar. Temperature: uniform. Items of interest: After an excellent supper our distinguished passengers smoked a final cigar and having sighted the French coast have now gone to bed to rest after the excitement of their leave taking. All essential services are functioning satisfactorily. The crew have settled down to watch keeping routine.'

At about 01.00hrs Sqn Ldr Johnston, after only a short break, returned to check the flight's progress. By chance a gap in the clouds allowed a quick and accurate position fix from the town of Poix, familiar to Johnston from previous journeys, and confirmed the relentless drift eastwards with R101 now 15 miles off the intended course from the French coast to Paris. Heading was adjusted to make the track good for Orly and the routeing by the Beauvais Ridge confirmed. In the journey log Irwin noted the following details, which are reproduced as accurately as the known facts permit:

Middle Watch: 23.00 to 02.00hrs
Officer of the Watch: H. C. Irwin
Time: 01.00hrs GMT
Duration from Start: 6hr 24min
Altitude: 1,500ft
Pressure Height: 1,500ft (This was the maximum height possible without venting gas.)
Temperature: 50°F
Track: 168° True
Wind Direction: 226° True
Wind Force: 49mph
Through Air Heading: 209° True
Engines: Five at 825rpm
True Air Speed: 54kt
Ground Speed Made Good: 30.15kt
Position: One mile northwest of Poix
Distance Flown: 193nm
Fuel used: 2.56 tons
Usable Fuel Remaining: 22.33 tons
Ballast Remaining: 6.5 tons
Cloud Base: Scattered, 1,000ft
Weather: Moderate rain

Just before 01.30hrs Johnston obtained cross bearings from Le Bourget (347° True) and Valenciennes (228° True) confirming position on track and was now satisfied that proper account was being taken of the wild and gusting winds. At their present ground speed R101 would be overhead Orly in just over 2¼ hours. Approaching Beauvais town, lying 30 miles northwest of Paris, the time was nearing 02.00hrs and the present crew would shortly be standing down from duty to hand over to the morning watch. A few minutes later R101 passed low over the town with many of the local inhabitants running from their homes to witness the spectacle. Flg Off Maurice Steff, one of the relatively inexperienced airship officers, arrived promptly to assume command from the fatigued Irwin who retired hastily to bed. Also sharing the morning watch, coming on duty in the control car, were Chief Coxswain Hunt, the Steering Coxswain Foster and Height Coxswain Mason. As the two 'helmsmen' acclimatised to the feel of the ship, crews in the engine cars, wireless room and elsewhere were completing the changeover. High in the framework another familiar figure was up and about, the brilliant assistant designer Michael Rope, conscientiously checking the bags and outer cover forward for signs of wear. Apart from spillage of gas from the sensitive side-mounted valves in turbulence there was also the problem of gas bag surging at the nose. A unique venting system provided an internal airflow through R101 to reduce the influence of outside pressure on the envelope and to purge any pockets of dangerous hydrogen/air mixture within. This internal airflow, however, gave rise to a bellows effect of the forward bags which additionally triggered valve operation. Any steady leakage of the front gas cells, especially to R101 which had essentially been running nose heavy since insertion of the new bay, could give cause for concern and was worthy of special attention.

In the control car meanwhile, all was as calm as could be expected under the circumstances, and details of the flight were duly being entered in the journey log book:

Morning Watch: 02.00 to 05.00hrs
Officer of the Watch: M. H. Steff
Time: 02.00hrs GMT
Duration from Start: 7hr 24min
Altitude: 1,200ft
Pressure Height: 1,500ft
Temperature: 50°F
Track: 108° True
Wind Direction: 225° True
Wind Force: 50mph
Through Air Heading: 209° True
Engines: Five at 825rpm
True Air Speed: 53kt
Ground Speed Made Good: 29kt
Position: Beauvais
Distance Flown: 215nm
Fuel used: 3.0 tons
Ballast remaining: 6.5 tons
Weather: Moderate rain
Usable Fuel Remaining: 21.88 tons
Cloud Base: 1,200ft

At a few minutes past 02.00hrs the scene was now set for the catastrophe about to unfold. In the dead of night, R101, battered by gusts and rain, strained with her fatigued and jaded crew to make headway against the strong winds. Flying at the cloud base height of 1,200ft so that sight of the ground could be maintained, the airship was flying at less than 500ft above the 774ft high Beauvais Ridge which lay to the southwest with its notorious downdraughts.

The series of incidents that led to the crash appear to have been preceded by some major failure of the outer cover and of one or both gas bags at the nose of the airship. It is likely that Sqn Ldr Rope, on his survey of the airframe, discovered a large tear near the highly stressed upper section of the nose in a portion of the outer cover which had not been renewed. With the airship being driven at high cruise speed any rent would quickly worsen and would result in the exposure and ripping of the front two gas bags. The sudden rupturing of the forward gas bags resulted in the airship being nose heavy, with gas continuing to escape rapidly. The nose pitched down sharply and the airship began to descend. All those awake were surprised by the severity of the action.

Left:
R101 on mooring post, 1930. *Courtesy Geoffrey Chamberlain*

If a downdraught existed in the lee of the Beauvais Ridge its effect, if any, would only have been marginal. Alarmed by the circumstances, the order was given to release ballast to arrest the rate of descent and from the control car valves were opened to discharge water ballast from the airship. At about this moment Sqn Ldr Rope probably arrived from his inspection with news of the serious damage at the nose and the danger of the situation. Immediately, Chief Coxswain Hunt rushed from the control car to order release of ballast forward and to warn the crew of their plight. On his way Rigger Church was met and dispatched the 300ft along the gangway to release one half-ton ballast tank at the nose. The height coxswain reacted swiftly and spun the wheel to up elevator. Further ballast was dropped and, with the powerful elevator selected to fully up, the airship began to respond. Level flight was maintained at about 500ft. The dive had lasted about 30sec, buy the airship was not to remain level for long. Flight could have been sustained in this condition but, with the airship appearing to be in great danger, the decision was taken to reduce power and the order to slow the engines was given. Whoever assumed responsibility for this course of action is not known but, with the reduction of rpm to slow power, the fate of R101 was sealed. At just after 02.07hrs the engine telegraphs were rung to indicate the lower power setting. The procedure to telegraph each engine car in turn would take some seconds and an even longer time would elapse before the engineers in the car reacted. As the forward engines were slowed, the loss of aerodynamic lift, compounded by the reduction in engine pitch up effect, resulted in a pronounced dive. In these circumstances Lord Thomson's carpet, all 129lb of it, stowed along the gangway right up by the nose, would not have helped.

By this time Hunt had reached the switch room en route to the crew's quarters.

'We're down lads. We're down lads,' he called.

There was no panic in his voice; no shouting or frenzy in his actions. It was simply a matter-of-fact declaration of their circumstances from a most experienced and able airshipman.

The height coxswain held the wheel to full up elevator but to no avail, and R101 dropped from the sky. A more experienced commander might have grasped the glaring danger of the situation. Reversal of the earlier decision and an order to apply full power immediately might just have allowed the airship to claw back some height. Calculations had shown that it was certainly capable of such a manoeuvre. Instead the airship continued inexorably toward its destruction. At just after 02.08hrs the aft engine was responding to the telegraph and reducing power. With all engines at 'slow' setting, R101's nose continued to dive with elevator still fully up and she dropped to her death. The airship struck the ground at an angle of about 12 degrees at precisely 02.09hrs in the lee of the Beauvais ridge, by the village of Allonne, flying in the strong head wind at a speed over the ground of only about 10-15mph. The impact was more of a crunch than a crash, but was sufficient to telescope the structure inwards by as much as 88ft. The control car collapsed under the strain and water from the drenched field soaked the box of highly inflammable calcium flares which had spilled on to the floor. Almost immediately a fierce fire erupted. A violent explosion followed as the hydrogen gas caught alight amidships and soon the airship was ablaze.

A Monsieur Woillery living on the northwest side of Beauvais, about 4.5km from the crash, woke his children to see the airship go by. They could see from the windows the lights of the ship passing to the east of the town, some distance beyond the cathedral. At 02.05hrs it disappeared behind another house and he sent his children to bed, but his 14-year old daughter went back to the window as soon as the others had gone and . . .

'I saw the lights of the ship reappear again beyond the house. I saw only a glimpse of the lights and then they appeared to be going down. The next instant the sky was lit up and a noise like a clap of thunder followed. When the explosion occurred I thought I could see the outline of the ship or at least part of it, and it was at an angle. The sky filled with pieces of burning wreckage which floated away, slowly sinking. The big flash lasted a few seconds and then steady illumination lit up high ground on the west side of town.'

In the smoking room aboard R101, the Cardington Foreman Engineer, Harry Leech, was finishing a final cigarette before bed when the airship went down.

'Within two seconds of striking, a blinding flash of fire occurred and this appeared to me to originate immediately above the control car. The door of the smoking room was burst open by the impact and it was through this opening that I saw the mass of flame. The next thing I knew was the the upper passenger deck had collapsed downwards on to a level with the tops of the backs of the settees in the smoking room, leaving me a space about 3ft high. This space was filled with choking fumes but not flames. I heard people screaming and moaning in the crew quarters and also from the upper passenger deck which was then blazing. I tore one of the settees away from the bulkhead and managed to scramble through the opening and found myself inside the hull on the starboard side. By that time the outer cover had been completely burned away from that section except for the Cellon windows which were still blazing and through which I had to force my way.'

In the No 4 engine car Arthur Cook had just come on duty at the change of the watch and was actually looking out of the nacelle door towards the ground when the impact occurred.

'My car sank on to the ground quite an appreciable time after the ship struck. In fact I had time to stop the engine. I remember pieces of structure falling on to and round the car. My first attempt to get out was unsuccessful and finally, after a struggle, I managed to get clear of the burning wreckage.'

In the aft No 5 car Engineers Binks and Bell were luckier than most. Joe Binks: 'A crash! Engine is stopped immediately. Explosion and fire. Aft car does not strike ground heavily but, in the ship's effort to telescope, the car is bumped along the ground causing the bottom to cave in; a blaze coming through the opening. "Looks as though all is up, Bell!" But no. Although by this time our car light must have gone out it was not missed owing to the fact of receiving plenty of illumination from the fire which was by this time all around the car. We endeavoured to quench the flames coming through the bottom of the car; getting too close to the petrol tank and further menacing our chance of escape; this tank for the starting motor now contained something like 12gal of petrol and should it become hot it might burst. However, it did not do any such thing even though it must have got fairly warm. The heat was certainly affecting both Bell and myself.

At this stage there was no chance or hope of escape through the exit of the car for we could see nothing but fire. There was much smoke also in the car and we begin to think we should be suffocated. Far better, Bell, than being roasted. Thought of many things now; family, friends; even discussed what a blow to Cardington Works and the general idea that R101 could hardly take fire. What a fallacy! Now, a sudden change and also our saviour. A deluge of water just when it was most needed, in our personal case; Bell and myself getting a goodly share, but I imagine the majority ran off the upper portion of the car and so subdued the flames in the immediate vicinity of the exits.

'We could now, thanks to this splendid and merciful water, put our heads out of the car exit and could see that much structure had either fallen on, or around it or, possibly, the action of striking the ground had forced us up and the strut girders from which the car was suspended were now resting on the car. One of the girders was athwartships and inclined downwards just clear of the exit.

'I felt sure that my colleague, Bell, like myself, must have been amazed at the transformation of our ship. Such pride and confidence we had in her and although not exactly being our home, our lives were very much wrapped up in her in many ways, apart from the fact that it was the means of earning our living.

'The exit being on the port side we could see straight across and through that side of the ship but we could not see what was beyond — and, personally, I didn't care, for now we were presented with a chance of getting clear of the fire so long as we kept our heads. And so, with wet rags held to our faces we left the car; Bell in the lead and in spite of various obstructions, we quickly arrived at the ship's side and, jumping or falling clear we landed on the fringe of a plantation of small trees.

'A few paces left and in the direction of the port quarter of the ship and we were now in rough grass, it was raining heavily and half a gale blowing diagonally across the ship from the starboard bow and bringing towards us thick smoke and flame from the accommodation parts of the ship.'

Arthur Bell:

'The first thing I knew after the car hit the ground was that there were flames all round us and, owing to the heat and fumes, we gave ourselves up for lost. A little later water started to run from the ballast tank above the engine nacelle and this doubtless saved our lives. We saw gaps in the flames and managed somehow to scramble out and run clear of the fire. Mr Leech and others joined us as we got clear.

'By that time the passenger accommodation was blazing but the bow and stern of the ship only showed patches of fire.'

In all only eight men escaped from the burning wreckage and two of these died later in hospital, leaving a total of six survivors out of a complement of 54. The accident was the second worst disaster in aviation history, the worst having occurred to a French Navy airship which was struck by lightning with the loss of 50 lives.

One of the survivors, Electrician Arthur Disley, called the Air Ministry from a nearby telephone with the information that R101 had gone down. News of the tragedy was soon transmitted to all stations down the line: to Malta, Cairo, Baghdad, Basra and Karachi. In the headquarters building of No 203 Flying Boat

Squadron, near Basra, Driver John Buchanan sat up late in the out of bounds wireless station with his friends, Operators Carruthers and Nash. A rush of morse on the headphones silenced the conversation.

'My God, it's the R101 — she's crashed.' Driver Buchanan grasped the note and ran to rouse his CO. 'She's down, sir', he gasped, 'the R101 has crashed.'

The Empire and all the world was stunned.

Those who perished were brought back across the Channel aboard HMS *Tempest* to lie in state on 10 October in Westminster Hall as the nation paid its respects. The next day they were taken by solemn procession through the streets of London to Euston station, and from there by special train to Bedford where gun carriages awaited. They were laid to rest together in the Church of St Mary within sight of the place from whence they departed: the Royal Airship Works, Cardington.

On 28 October 1930 the inquiry was opened by Sir John Simon, appearing on behalf of the Crown, with these words spoken to the assembly:

'I think it would be fitting if we stood for a moment to express our sense of the poignancy of the tragedy which we are met to investigate, and our sympathy with all those to whom the dead were near and dear.'

All stood in silence for a few moments.

The proceedings which ensued were somewhat deficient when compared with modern investigations. The route taken by R101 was not even examined. The commission, poorly versed in the art of accident investigation, simply concluded that substantial loss of gas in bumpy conditions, compounded by a downdraught, was the probable cause. It was a totally inadequate explanation for the sequence of events known to have taken place.

In the end the airship programme was cancelled. R100 was broken up in its shed, and the wreckage of R101 brought back to London where it was sold at auction for £440. The grand plans for R102 and R103 were abandoned. Had today's rigorous standards of safety been applied in the past to airship development, who knows what a gentle and gracious travelling style these mammoths might have brought to the air. Aircraft design, however, was proceeding fast, and had an Empire airship service been introduced it would probably have been short-lived, although, of course, airships may have proved useful in other fields. Today airship development continues in limited programmes with smaller craft, and airships may yet find their place. Whether the demise of the airship industry with the crash of the R101 was a blessing in disguise, or the unfortunate loss of a useful airborne craft, is the source of much interest and speculation, and is an argument which is still open to this day.

List of those on board R101

(The names of survivors are in italics.)

Passengers:
Brig-Gen The Rt Hon Lord Thomson PC, CBE, DSO (His Majesty's Secretary of State for Air)
Sir W. Sefton Brancker KCB, AFC (Director of Civil Aviation)
Maj P. Bishop OBE (Chief Inspector, AID)
Sqn Ldr W. Palstra (Representing Australian Government)

Sqn Ldr O'Neill (Deputy Director of Civil Aviation, India, Representing Indian
   Government)
Mr James Buck (Valet to Secretary of State for Air)

Officials from the Royal Airship Works:
Wg Cdr R. B. B. Colmore OBE (Director of Airship Development)
Maj G. H. Scott CBE, AFC (Assistant Director (Flying) Officer in Charge of
   Flight)
Lt-Col V. C. Richmond OBE (Assistant Director (Technical))
Sqn Ldr F. M. Rope (Assistant to Assistant Director (Technical))
Mr A. Bushfield (Aeronautical Inspection Directorate)
Mr H. J. Leech (Foreman Engineer)

R101 Officers:
Flt Lt H. Carmichael Irwin AFC (Captain)
Sqn Ldr E. L. Johnston OBE, AFC (Navigator)
Lt Cdr N. G. Atherstone AFC (1st Officer)
Flg Off M. H. Steff (2nd Officer)
Mr M. A. Giblett MSc (Meteorological Officer)

R101 Crew:
Chief Coxswain: G. W. Hunt
Asst Coxswains: Flt Sgt W. A. Potter, I. F. Oughton, C. H. Mason
Riggers: E. G. Rudd, M. G. Rampton, H. E. Ford, C. E. Taylor, A. W. J.
   Norcott, A. J. Richardson, P. A. Foster, W. G. Radcliffe (died at Beauvais
   6/10/30, S. Church (died at Beauvais 8/10/30)
1st Engineer: W. R. Gent
Charge-hand Engineers: G. W. Short, S. E. Scott, T. Key
Engineers: R. Blake, C. A. Burton, C. J. Fergusson, A. C. Hasting, W. H. King,
   M. F. Littlekit, W. Moule, A. H. Watkins, *A. V. Bell, J. H. Binks, A. J.
   Cook, V. Savory*
Chief Wireless Operators: S. T. Keeley
Wireless Operator: G. H. Atkins, F. Elliott, *A. Disley*
Chief Steward: A. H. Savidge
Stewards: F. Hodnett, E. A. Graham
Galley Boy: J. W. Megginson

Flights carried out by R101

No of Flight	Date	Itinerary, etc	Duration of Flight
			Hr Min
	12/10/29	Ship brought out of shed	
1	14/10/29	Round London	5 38
2	18/10/29	Midlands	9 38
	21/10/29	Ship in shed	
	1/11/29	Ship brought out of shed	
3	1/11/29	Norfolk	7 15

Flights carried out by R101 (continued)

No of Flight	Date	Itinerary, etc	Duration of Flight Hr	Min
4	2-3/11/29	Isle of Wight	14	2
5	8/11/29	Local	3	4
6	14/11/29	Local	3	9
7	17-18/11/29 (Endurance Flight)	England, Scotland and Ireland	30	41
	30/11/29	Ship in shed		
	23/06/30	Ship brought out of shed		
8	26/06/30	Refit Flight	4	35
9	27/06/30	Rehearsal for RAF Display	12	33
10	28/06/30	RAF Display Flight	12	21
	29/06/30	Ship in shed. New bay inserted		
	1/10/30	Ship brought out of shed		
11	1-2/10/30	First Trial Flight with extra bay	16	51
12	4-5/10/30	Cardington-Beauvais	7	24
		Total Flying Time	127	11

Below:
R101 crash site near the village of Allonne. *Courtesy Geoffrey Chamberlain*

The Comet
Crashes   1953-54

At the end of World War 2 the de Havilland Aircraft Co took the courageous decision to begin construction of the first ever jet airliner. The experience they had gained building Vampire jet fighters in wartime was to be put to good use in a revolutionary venture in peace. In all other aspects of commercial operation the Americans had secured an unassailable lead from their domestic operations during the war and de Havilland's ambitious plans were designed to place Britain back in the forefront of aviation technology.

In the United States new types of airliner, like the pressurised Douglas DC-6 with the advanced reversible-pitch propeller, were being introduced into service but all were designed along conventional lines with only improvements to existing models. Competing against American manufacturers in this field was futile under the circumstances and the future of British aviation seemed to lie in extending into the unknown. In one field in particular, jet engine operation, European development, with Sir Frank Whittle's work in Britain and Heinkel's in Germany, was ahead of American progress. While the American P-84 jet fighter was establishing a national speed record of 611mph in California in 1946, the British were already experimenting with a Lancastrian with two of its propeller engines replaced by jets. The resulting hybrid ably demonstrated jet engine performance by flying 100mph faster than any other Lancastrian.

Credit to the de Havilland Co for its vision and bravery in embarking upon such an adventurous programme so early in jet development has been somewhat overshadowed by subsequent events. The proposed jetliner was to fly almost twice as fast and high as existing machines and would lead the company into realms of air travel unexplored at that time. Jet engines were known to operate efficiently at high rpm and only at high altitude could high rpm be set without producing excessive thrust. Engine performance, however, was known to diminish with increasing altitude as the air becomes thinner, but the same was also known of drag, the force of the air resisting aircraft movement. Loss of performance, therefore, in the operation of the jet engine at great height would be more than overcome by a combination of reduction in aircraft drag and better engine efficiency.

Jet airliners would be compelled to fly fast and high which would, of course, be their great advantage over conventional types, but the problems to be overcome posed enormous difficulties for the design teams. Flying at speeds of around 500mph through the air an entirely new aerodynamic concept would be required with fine lines and sweptback wings. Air loads on the control surfaces would require development of completely new flying control systems and new aircraft

36

and engine handling techniques would be required. Cruising at heights of up to 40,000ft would require effective cabin pressurisation, and new metal bonding techniques would be needed for strong, lightweight aircraft construction. Pressurisation of aircraft hulls was not new, of course, but had not been applied to the extent of this requirement. The fuselage would have to contain the pressurised air of the cabin and prevent it from bursting outwards into the rarefied atmosphere, and the structure would be subjected to a stress not experienced in the past.

As early as 1942, with the British nation sensing some hope of victory in the war, a committee was set up under the chairmanship of Lord Brabazon of Tara to formulate a national peacetime aviation policy. One of the more ambitious projects, the Brabazon IV, called for the development of a jet mailplane for North Atlantic service. The aircraft payload was to be one ton with no passengers. By 1944 the concept had changed to a passenger-carrying jetliner for European operations, but with the potential for development of North Atlantic flights. Later that year the British Overseas Airways Corporation's (BOAC) indicated requirement for 25 Brabazon IV types inspired the successful de Havilland Co to commence development of the jet airliner. The original Brabazon project specification suggested a 14-seat aircraft of 800 miles range, but de Havilland were already thinking of a 24-passenger version using four 5,000lb thrust Halford H2 centrifugal flow turbojets of their own manufacture, to be known as the de Havilland Ghost jet engine. As early as 1945 project DH106 was begun in great secrecy under de Havilland's chief designer, Ronald Bishop. A tailless aeroplane was considered and an experimental machine, the DH108 with a 40° sweptback wing, was constructed using a Vampire fuselage. During a trial flight on 27 September of that year the prototype broke up in a steep dive killing the pilot, Geoffrey de Havilland junior. Control modifications were introduced but proved unsatisfactory and another two experimental aircraft crashed, also killing their pilots. Some success was achieved when one trial craft broke the sound barrier but the problems of stability and control were taking too long to solve and the idea was finally abandoned. After failures of the tailless DH108, design of the DH106 project settled on more conventional lines. By the end of 1946 the standard design was complete: the wing sweep had been reduced from 40° to 20°, thus increasing payload but decreasing cruising speed. The final plan for a 32-seat, four-jet aircraft of around 1,600-mile range was conceived. Impressed by the results the Ministry of Supply ordered two straight from the drawing board. The aircraft was just what was needed for the flagging British aviation industry. BOAC was equally impressed with the passenger appeal and ordered 14. At £450,000 each a delighted de Havilland had £7 million worth of orders even before one had been built.

Towards the end of 1947 development of the DH106 prototype, named the Comet, continued in great secrecy, and it was not until April 1949 that the aircraft was revealed to the public. On 7 July the Comet prototype G-ALVG (known as Victor George from the old phonetic alphabet) took off from Hatfield on its maiden flight and caused a sensation throughout the world. Its sleek lines pleased the eye and its statistics impressed the airlines. Already, larger and more powerful variants of the Comet were on the drawing board with increased payload and range. The Americans were years behind and were caught completely off balance. The British, for once, had the field to themselves.

Extensive flight testing of the aircraft continued throughout the summer months with satisfactory performance. In September of 1949 the Comet was put through its paces at the Farnborough Air Display by chief test pilot John Cunningham, and the orders poured in from throughout the world. A year later a second Comet prototype was completed. The two aircraft successfully flew a whole series of tests, shattering speed records wherever they went.

The first production aircraft, the Comet 1, flew in January 1951 with the only major change from the prototypes being the conversion from a single-wheel main undercarriage to a four-wheel bogie. The Comet 1 could fly 36 to 44 passengers at 490mph (cruising Mach number 0.74) over a 1,750-mile range with maximum payload, or 3,860 miles with maximum fuel. Its maximum take-off weight was 105,000lb, with a total fuel capacity of 40,000lb carried in integral wing tanks. A new light, thin aluminium alloy was used in the construction and comprehensive use was made of metal-to-metal bonding. The skin of the fuselage and wings was secured to framework stringers using Redux plastic glue, an extremely powerful metal adhesive. At an aircraft height of 40,000ft, equivalent to 7½ miles above sea level, a cabin altitude of only 8,000ft resulted in a differential pressure between inside and outside the fuselage of about 8½lb/sq in being exerted on the structure. The machine had been tested in conditions of severity hardly likely to be encountered in flight, and the cabin had been subjected to pressures far beyond international regulation requirements.

With success at de Havilland's fingertips plans were laid to increase Comet production. In February 1951 the experimental Comet 2X, powered by Rolls-Royce Avon axial flow engines, took to the air. The extended fuselage allowed higher payload, and improved fuel consumption increased range. The Comet 2 looked promising.

On 2 May 1952 the Ghost-engined Comet 1 G-ALYP (Yoke Peter), under the command of Captain Majendie, entered service on the London-Johannesburg route. Scheduled refuelling stops were arranged for Rome, Beirut, Khartoum, Entebbe and Livingstone. A large crowd assembled at London Airport to witness the BOAC Comet's take-off inaugurating the world's first regular passenger jet service. It was a great achievement for British aviation.

The Comet basked in summer glory. Princess Margaret and the Queen Mother became its first royal travellers. The aircraft was popular with passengers and BOAC became the envy of other airlines. In September of that year the transatlantic Comet 3, having a 2,700-mile range with maximum payload, was announced. The maximum take-off weight was to be 145,000lb, with the fuselage further stretched to carry 58 to 76 passengers. Four Rolls-Royce Avon RA26s, each of 9,000lb static thrust, were to be the powerplants. The Ministry of Supply ordered a prototype Comet 3 and BOAC and Air-India placed orders. In October, Pan American, devoid of any equivalent American machine, joined the list of Comet 3 buyers by ordering three and taking options on seven more.

By 26 October, just less than six months since the inaugural service, the first Comet was lost. The northbound BOAC service from Johannesburg to London crashed on take-off with 33 passengers on board after a scheduled stop at Rome's

Ciampino airport. On a wet evening the aircraft shuddered at lift off and the captain made the somewhat late decision to abandon take-off. The aircraft careered off the runway, striking an earth mound which tore off the undercarriage, before coming to rest. Fortunately the broken aircraft did not catch fire and everyone evacuated safely. Pilot error was found to be the cause of the near disaster and the captain was demoted to flying York freighters. A landing accident then occurred at Entebbe, Uganda, in January 1953 when a Comet 1 undershot the runway killing an airport worker, the only casualty.

On 3 March of the same year another Comet 1 crashed on take-off, this time at Karachi. The aircraft, *The Empress of Hawaii*, was on a Canadian Pacific Airline's delivery flight to Sydney, Australia, where it was to open up a new Pacific service via Honolulu. In the half light of early dawn the aircraft failed to become airborne, crashed through a fence and skidded into a high bank where it burst into flames. The brand-new Comet was completely destroyed and the crew of five, plus six technicians aboard were all killed. Once again pilot error was blamed: the captain had little experience of jet flying or of operation in hot climates. Following the crash, Canadian Pacific cancelled their orders for another two aircraft.

The Comet 1 was known to leave little room for error during take-off and landing, but the two crashes, plus a number of other related incidents, were beginning to prove that there was more than pilot handling at fault. It was known, for example, that if the Comet 1's nose was pulled up too quickly at lift-off the wing, because of its swept back characteristics, could stall before the aircraft left the ground. Also, the control system on the early Comets did not help the delicacy of the manoeuvre. Direct physical movement of control surfaces on passenger jets is beyond human strength and the Comet controls were hydraulically operated. To overcome the lack of sensation of aerodynamic loads on the pilot's controls a crude artificial spring-loaded 'feel' system was introduced on the Comet 1. The force felt by the pilot on the spring-loaded control column, therefore, was related simply to column position, and not to airspeed. At take-off it was all too easy to pitch up sharply, and at night judging the correct nose up attitude was difficult. Later Comets incorporated what was known then as 'Q' feel where feel units introduced loads to the pilot's controls related to both control surface deflection and airspeed. Wing leading edges were also modified to alleviate the stalling problem. As a note, the captain blamed for the Rome take-off incident was eventually completely exonerated.

On 2 May 1953, the first anniversary of commercial jet flights, a major Comet disaster occurred with the loss of 43 lives. Comet 1 G-ALYV (Yoke Victor) flying BOAC's westbound Singapore to London service with Captain M. Haddon in command, broke up in flight en route to Delhi in a violent tropical storm of exceptional severity. Six minutes out of Calcutta's Dum Dum Airport, passing through about 10,000ft, Yoke Victor radioed 'climbing on track' just before going down. Eyewitnesses from Jugalgari village, 30 miles northwest of Calcutta in the Hooghly district, reported that 'an aeroplane has been knocked down by the tempest'. They told of seeing a red flash in the sky as the aircraft caught fire during the intense electrical storm. What appeared to be a wingless aircraft plummeted towards the ground, flashing and exploding as it passed low over the trees, before crashing in a mass of flames. The wreckage was scattered over eight square miles with the wings and engines lying four miles apart.

An Indian inquiry established that the port elevator spar had failed followed by failure of the wings at rib number seven as the aircraft plummeted out of control. It was noted that during development, metal fatigue tests conducted on the wing had resulted in failure at the same location, but modifications had been carried out. Excessive loads experienced beyond design limits in the dive would, of course, have caused failure of the section. Metal fatigue, however, is a different consideration, and is the weakening effect of frequent manipulation, or application of stress, over a length of time, which eventually reduces component strength to a point where failure occurs under normal load. Obviously aircraft should be withdrawn from service before this point is reached allowing safe operation during the life of the machine. The early detection of metal fatigue is of paramount importance.

The Indian inquiry, however, concluded that Yoke Victor, while climbing through the thunderstorm, had encountered a severe squall of gust velocities so high that any aircraft might have been endangered. It was also suggested that the pilot might have overcorrected in the exceptional turbulence resulting in failure of the elevator spar. In the wreckage the throttles were found in the half open position, indicating they were retarded to reduce speed in a dive. The reaction of the pilot in attempting to pull out of the descent may have overstressed the elevator owing to the crude control feel system. It was recommended by the inquiry that the feel units be modified. Wreckage was also to be sent to the Royal Aircraft Establishment (RAE), Farnborough, for analysis, although the inquiry's suggestion of control difficulties was already diverting attention from any fatigue problems that may have existed. The RAE had extensive experience of analysing flight failures, however, and could be expected to complete a most thorough examination. Even as early as 1930 they had conducted the first investigation into the break-up of an all-metal aircraft when the wreckage of a Junkers F13, which had broken up in the air, came into their hands. The aircraft, G-AAZK of Walcot Airlines, disintegrated in cloud at Meopham, Kent, on 21 July 1930. The tailplane failed in turbulence resulting in break-up of the tail unit assembly followed by failure of the port main wing and disintegration of the fuselage. All six on board were killed. The report of the investigation recommended that (1) The phenomenon described as 'buffeting' should be investigated, (2) Steps should be taken to place the measurement of bumpiness on a scientific basis, (3) Removal of parts from wreckage by souvenir hunters should be discouraged!

In the end, insufficient wreckage of the Comet was sent from Calcutta to Farnborough. When more was requested it was discovered that the Indians had irrevocably disposed of the remainder. Even if all the remnants of Yoke Victor had been available to the RAE investigators it is impossible to say whether early evidence of excessive metal fatigue would have been discovered. It may have been chanced upon during inspection although even then it may not have been accorded much importance. The Indian inquiry's finding that the accident resulted from overstressing of the controls was never disputed and was generally accepted as the probable cause. Evidence of metal fatigue of the wings was never substantiated and the suggestion quickly receded into the background.

The British Air Registration Board (ARB) was already calling for more strenuous static loading tests of civilian aircraft, in line with RAF procedure, and de Havilland was in the process of performing additional experiments. Sections of the cabin were subjected to fatigue loading tests far in excess of ARB requirements in an effort to determine the Comet's safe working life. The tests ended in the autumn of 1953. The fuselage skin was shown to suffer a fatigue failure at the corner of a window after repeated applications of excessive loads. The trials, however, were considered so extreme that the researchers deemed it inconceivable that failure of that nature could occur during the normal working life of an aircraft. The next year was to prove the scientists tragically wrong.

De Havilland and BOAC, both deeply concerned with the Comet 1's image, issued a joint statement. Until the accident to the Calcutta Comet was more fully examined it was possible only to theorise on Yoke Victor's break-up. The Comets continued to fly, but before long another aircraft crashed. A French Comet 1 was destroyed when it overshot the runway at Dakar, thus bringing the tally of serious accidents to five in the 18 months since commercial Comet operations began. The Comet 1 accident record was proving to be disturbingly bad.

On Sunday 10 January 1954 BOAC's Comet 1 service from Singapore to London (BA781) departed Rome at 09.31hrs GMT (10.31hrs local) on the final leg home. The flight was an extra westbound service which had also made scheduled refuelling stops at Rangoon, Calcutta, Karachi and Beirut. The morning was typical of a fine Mediterranean winter's day, with sunny weather and only thin broken layers of cloud. Aloft, a strong wind blowing like a fast flowing river of air, known as a jetstream, created some clear air turbulence, but otherwise flying conditions were perfect. The Comet 1 G-ALYP (Yoke Peter), the same aircraft which had flown the inaugural service the previous year, climbed rapidly to its cruising altitude of 36,000ft. Under the command of Captain Alan Gibson the flight proceeded northwest up the coast of Italy, passing radio messages as the journey progressed. Seven transmissions were sent in quick succession, including two position reports. At 09.34hrs GMT, Yoke Peter called passing the radio beacon at Ostia on the coast, climbing through 6,500ft. Again at Civitavecchia, the aircraft reported 19,000ft, climbing, at 09.42hrs. Nineteen minutes after departure, 09.50hrs GMT, Yoke Peter reported the Orbetello beacon, passing 26,000ft, clear of the cloud and climbing to 36,000ft. The aircraft then turned toward the island of Elba, leaving the coast of the Italian mainland behind.

Before departure Captain Gibson had conversed with the captain of a BOAC Argonaut aircraft. The Argonaut had taken off 10min before Yoke Peter, but before leaving had asked for reports of cloud heights. The Comet 1 was to pass the company aircraft on its way and would arrive in London two hours ahead. Captain J. Johnson of the Argonaut noted the cloud base to be 10,000ft, and radioed Yoke Peter requesting the height of the tops. At 09.52hrs the Comet transmitted the requested information with no response from the Argonaut. Yoke Peter made another attempt at contact.

'George How Jig from George Yoke Peter, did you get my . . .?'

The Comet 1 disintegrated in flight at about 27,000ft, shattering into a thousand pieces which fell over a wide area. Twenty-nine passengers and six crew perished.

Below:
G-ALYP flight plan, 10 January 1954.

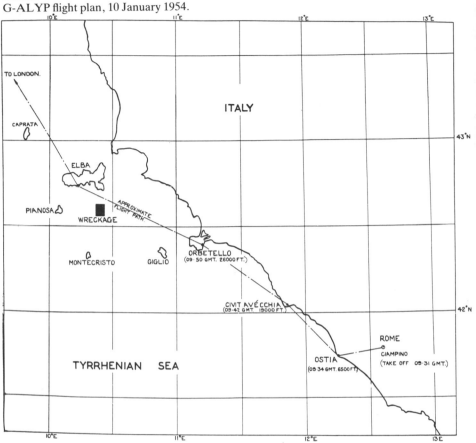

Among the dead were 10 children returning to school in Britain after visiting parents abroad during the Christmas break.

An Elba fisherman, Giovanni di Marco, at sea with his partner when the catastrophe struck, was the first to inform the Carabinieri at Porto Ferrajo on Elba of the crash. Reports were then relayed to the authorities in Pisa. The two men had been fishing south of the island when they heard, unseen above the clouds, the whine of an aircraft over their heads. Three explosions were heard in quick succession, then silence. Suddenly, several miles away, di Marco 'saw a silver thing flash out of the clouds. Smoke came from it. It hit the sea. There was a great cloud of water'. The fishermen rushed to the scene but there was nothing they could do. They picked up some bodies lying in the water which had stilled again after the crash, then returned to Elba to raise the alarm. In the hills to the north east of Elba, behind Porto Ferrajo, a farmer out shooting reported being 'on a small hill where the gun emplacement used to be. I heard the noise of an aircraft but didn't pay much attention. Suddenly I heard a roaring noise and I distinctly noticed, in the direction the noise came from, two pieces of an aircraft,

42

the smaller in flames, falling in almost parallel lines into the sea'. Another witness also spoke of hearing 'a roaring noise, like thunder. I turned in the direction of the noise which came from out to sea. I saw a globe of fire rotating as it came down into the sea. I saw it plunge into the sea, leaving a cloud of smoke over the air'. A lorry driver told of a number of explosions and a great roaring noise being heard. He 'turned and saw flames falling into the sea followed by a wake of smoke in the form of a spiral'.

The Italians, to their credit, quickly launched a rescue operation, but to no avail. A small armada of fishing vessels, joined by three Italian ships and search aircraft from La Spezia and Pisa, failed to find any survivors. Fifteen bodies were picked up. By nightfall, a few seat cushions, a scattering of personal effects, some mail bags, and only fragments of wreckage were all that was found.

News of the disaster shocked the British public. The beautiful, revolutionary new Comet, poised for world success, now seemed to be falling from grace as well as from the sky. A market of enormous potential was slipping from the British grasp. No other country, including America, was in the position to build another jetliner, yet potential customers were beginning to have doubts about the de Havilland machine.

The next day, Monday 11 January, BOAC made the following announcement:

'As a measure of prudence the normal Comet passenger services are being temporarily suspended to enable minute and unhurried technical examination of every aircraft in the Comet fleet to be carried out at London Airport.'

'Sir Miles Thomas (BOAC's chairman) has decided to devote himself almost exclusively to probing the Comet mishap, with Sir Geoffrey de Havilland and the highest authorities in Britain. His decision is based on a desire to retain the good name of the Comets.'

The Certificate of Airworthiness remained in force and BOAC's grounding of the Comets was entirely voluntary. With a mixture of fear and hope the suggestion of sabotage was advanced. Elements in both Beirut and Rome were certainly capable of placing explosives aboard an aircraft. Sir Miles stated that 'the possibility of sabotage cannot be overlooked. Special security investigators are being sent to points along the route to the Far East'.

The Italian commission, specially convened to examine the evidence, took the wise decision a few days after the tragedy to hand responsibility for accident investigation to the State of Registry. British experience in manufacturing and operating the aircraft would be invaluable in an inquiry and it was the most sensible course. A committee was quickly formed under Mr C. Abell, the BOAC Deputy Operations Director (Engineering), to contemplate modifications necessary before flying the Comets again, and 'to consider what possible feature or combination of features might have caused the accident'.

The small amount of wreckage found at the crash site was sent to the RAE for analysis and the Royal Navy was called in to help in the salvage operations. Extensive searches of the sea bed were begun, but working in difficult conditions only small amounts of wreckage were brought to the surface. With little concrete evidence on which to work the Abell Committee probed in the dark for likely malfunctions. Control flutter, hydraulic malfunction, loss of control, and catastrophic engine failure were among the items considered. Analogies were drawn between the Yoke Victor crash at Calcutta and Yoke Peter at Elba in the hope of finding a common cause not yet contemplated. Both aircraft had similarly broken up in flight and dropped in flames. Could some inherent weakness of the elevator spar or the wing still lie hidden in the depths of the structure in spite of extensive testing? The Calcutta incident, however, had occurred at about 10,000ft in a severe thunderstorm while at Elba the aircraft had broken up at about 27,000ft in what was possibly clear air turbulence, but not of a degree that was considered excessive.

In Italy, meanwhile, pathologists conducted autopsies on the bodies recovered from the crash site. Fire in the cabin as the result of bomb detonation, or explosive engine disintegration, would have inflicted severe burning on the occupants, but the conclusions reached by the Italian medical examinations contradicted the sabotage or catastrophic engine failure theories. In the Yoke Peter case, although burning was evident, the doctors demonstrated quite conclusively that the scorching had occurred *after* death. Injuries sustained by those on board indicated that they had suffered explosive decompression, with death resulting from the occupants being thrown violently, probably forwards and upwards, against the aircraft structure. The Italian pathologists' report, an incredible deduction at such an early stage of the investigation, clearly inferred break-up of the cabin. But how was this possible? Yoke Peter had completed only 3,681 flying hours since construction and its fuselage and wings had been subjected to exhaustive tests before operation. Its cabin had been built to withstand pressure loads two and a half times normal. Earlier stringent trials had shown the pressurised fuselage skin to fail near the corner of a cabin window, but the fracture had occurred only under repeated massive loads. Metal fatigue would, of course, reduce cabin strength but could hardly be considered at such an early stage of this aircraft's life.

Off the south Elba coast the Royal Navy search for wreckage continued. An impressive fleet of ships had now assembled for the task including HMS *Barhill* and salvage vessel *Sea Salvor* from Malta, two deep sea trawlers, *Ravilla* and *Carmelina*, chartered from Anglo-Mediterranean Salvage Company, and HMSS *Wrangler* and *Sursay* from England. Using grabs, pick ups, observation chambers, and for the first time, underwater TV cameras, the Naval teams fished desperately for weeks in the deep waters with only limited success. Debris lay on the sea bed at depths between 450 and 600ft, and might be scattered anywhere within a wide area. The harbour master at Porto Ferrajo, after interviewing witnesses on Elba, Piarossa, Monte Cristo, and the mainland, deduced that pieces of the Comet 1 could have fallen in an area as large as one hundred square miles. It seemed a hopeless task. By good fortune, however, a photograph taken from the air just after the crash proved useful and helped to concentrate the search. A crew member aboard a Skyways flight en route to the Middle East had taken a shot of the fishing boats picking up bodies and wreckage. The photograph also showed a corner of Elba from which it was possible to establish an estimate of position.

Storms lashed the area hampering the search and mud churned up from the bottom rendered TV cameras useless. Asdic equipment aboard the antisubmarine frigate was used to pin-point fragments while the trawlers dragged the sea bed and other vessels marked the lanes already searched with buoys. In the middle of February, just over a month after the tragedy, the asdic detected some substantial pieces of wreckage. At last the TV cameras sighted the parts, and at the end of February/beginning of March the first significant pieces of Yoke Peter were retrieved. A large section from the rear of the aircraft was brought to the surface by *Sea Salvor's* large toothed grab. The monotonous work was beginning to pay dividends. Soon much of the rear half of the fuselage and tail followed, including the circular pressure dome which sealed off the aft end of the cabin. Also found were some seats, the toilets and the bar, complete with bottles intact.

Further storms drove most of the salvage fleet into harbour except the larger *Sea Salvor* which remained at her moorings. On 15 March, during a lull in the bad weather, a diver lowered in the observation chamber spotted what proved to be the rear spar, the strong cross-member that runs through the wings and the fuselage. It lay only half a mile from the already retrieved tail section. At 60ft long and 2ft deep it was the largest section salvaged. Nearby the front spar, two engines and the centre section, the attachment for the wings, were also found. Later the other two engines and the undercarriage were brought to the surface. The findings were better than had been hoped, but much of the Comet had yet to be recovered.

In England, in the nine weeks since the crash, extensive examinations of the grounded Comets had shed little light on the problem. All possible causes of the accident were analysed and a total of 50 major and minor alterations were recommended by the Abell Committee and carried out. The chief modification was the installation of armour plating around the engines. Any catastrophic engine failure could then be safely contained and would prevent broken turbine blades from penetrating fuel tanks or cabin.

Without further substantial amounts of wreckage being found off Elba, a most unlikely occurrence, little more could be done to substantiate a reason for the accident. Expert opinion was at a loss as to the cause and the conclusion was finally drawn that the crash could have been the result of any number of factors. It appeared nothing more could be gained at the time from the further researches and keeping the Comets grounded seemed pointless.

From the outset, the Ministry of Transport and Civil Aviation's Accident Investigation Branch (AIB) had been charged with the responsibility of conducting the investigation. AIB representatives directed the search off Elba, examined wreckage on site, and arranged transportation of the pieces to Farnborough. Although not members of the Abell Committee they had been in attendance at all the meetings and were well versed with the course of events. Since no public inquiry had been ordered by the government, the Chief Inspector of Accidents of the AIB would be the person responsible for the production of a formal report. The Abell Committee, in agreement with the Air Registration Board (ARB) and the Air Safety Board (ASB), felt it unnecessary to delay services until the publication of the formal report. Nothing new would be added to the findings. BOAC had voluntarily grounded the Comet 1 fleet and was losing £50,000 a week as a consequence, so did not want to keep the aircraft on the ground needlessly. On 23 March 1954 therefore, 10 weeks after the tragedy and with the airline half a million pounds out of pocket, the Comet 1 scheduled services resumed with the Minister's consent. Public opinion had settled on the sabotage theory for the Elba accident and severe turbulence for the crash at Calcutta. Confidence in the aircraft's performance had not been impaired and the first flight after grounding departed for Johannesburg only two short of a full load.

Off Elba the search for fragments of Yoke Peter continued with increasing success. On 4 April, while the main pieces retrieved earlier were just arriving at the Royal Aircraft Establishment (RAE), a significant section of the front fuselage, forward of the wings right up to the nose, including the entire flight deck, was found. It was a major discovery.

For two weeks the Comet services continued without mishap and the decision to recommence flights seemed well founded. BOAC's confidence in the machine, as well as that of its rivals, appeared justified. The airline's competitor on the London-Johannesburg route, South African Airways, chartered a Comet 1 for

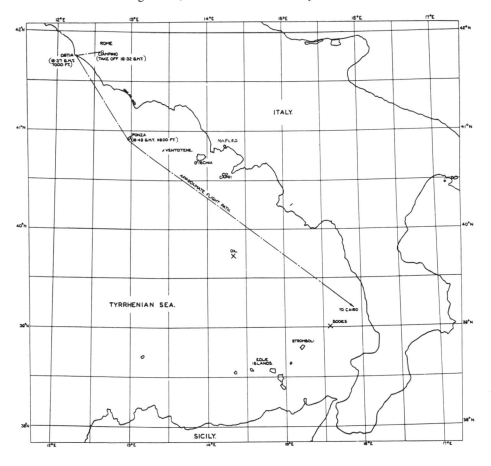

Above:
G-ALYY flight plan, 8 April 1954.

46

operation of its own services. On 7 April 1954, therefore, at 15.00hrs GMT, Comet 1 G-ALYY (Yoke Yoke), operating flight SA201, departed London for Rome with a South African crew on its first leg southbound to Johannesburg.

Flight SA201 arrived safely in Rome some 2½ hours later at 17.35hrs. During routine transit checks at Ciampino an engineer discovered that 30 bolts in a wing panel of the left wing had worked loose. The fault lay at London Airport where only a few days before the panel had been removed and had been incorrectly replaced. There was also a problem with a fuel contents gauge which would require fixing before continuing. Work to clear the faults delayed departure for 25 hours, but by the evening of Thursday 8 April the aircraft was ready for the flight. The next stop was Cairo.

Under the command of Captain W. Mostert, Yoke Yoke took off at 18.32hrs GMT with 14 passengers and seven crew, and climbed rapidly in the dark towards its cruising height of 35,000ft. The flight reported over the Ostia beacon at 18.37hrs, passing 7,000ft. The weather was good with no turbulence forecast, but the sky was overcast with layers of cloud. Another transmission was made at Ponza at 18.49hrs, climbing through 11,600ft. At 18.57hrs the aircraft reported passing abeam Naples, and at 19.07hrs, still climbing, Cairo was contacted on high frequency, long-range radio with an estimated arrival time of 21.20hrs GMT. Afterwards there was silence. The aircraft disintegrated unseen in the night sky, at about 30,000ft, killing all on board. Yoke Yoke was never heard from again. It had completed only 2,704 flying hours since construction. Repeated attempts by both Rome and Cairo to re-establish contact were made without success and the alarm was raised. A German radio station eavesdropping on the frequency picked up the transmissions and informed the press. Soon it was announced to the world that another Comet 1 had been tragically lost.

BOAC immediately, and once again voluntarily, grounded the Comets. The Italian air-sea rescue services were notified and at dawn were mobilised into action. The Royal Navy was alerted and the carrier HMS *Eagle* and the destroyer HMS *Daring* were dispatched early on the morning of the 9th to the estimated scene of the crash south of Naples. Fighters launched from the carrier, with Italian, British, American and Scandinavian aircraft, aided by the two British Naval ships and Italian vessels, all swept the area without success. Later a report was received from a British European Airways Elizabethan aircraft of a large patch of oil lying 70 miles from Naples and wreckage and bodies being sighted 30 miles northeast of Stromboli. The search was conducted further south. By dusk, five bodies and some flotsam were recovered and taken aboard HMS *Eagle*. A sixth body was found washed ashore. The remains of Yoke Yoke had sunk in water of depths in excess of 3,000ft and were lost forever.

One hundred and eleven lives had been lost since the Comet first flew and the aircraft now had a worse safety record than any other airliner in service. That evening the Minister, Mr Lennox-Boyd, announced removal of the Comet 1's Certificate of Airworthiness and the aircraft was officially refused permission to fly. Comets were stranded in London, Colombo and Cairo. On the South African route all BOAC's Comets were fully booked for the next three months. The crashes were a crippling blow to the dedicated men of de Havilland, and with cancellations the company stood to lose £40 million worth of cancelled orders.

Unless a solution to the problem was found quickly and the fault cleared, the grounding of the Comets could prove a disaster for British aviation.

Of the three aircraft lost in flight, one had broken up at a relatively low height in severe turbulence, while the other two had been lost in remarkably similar circumstances. Both had disintegrated at high altitude within 30 miles of Rome. If sabotage was suspect, Rome's Ciampino Airport, not renowned for its strict security, could be the linking factor. Examination of the bodies retrieved from the sea south of Naples after the Yoke Yoke crash, however, displayed injuries exactly the same as those of Yoke Peter's occupants recovered off Elba. Once again, internal injuries sustained by the victims indicated explosive decompression. With an instantaneous loss of pressurisation, air in the body would rapidly expand, and all those killed showed signs of severe damage to the lungs. If the medical reports were to be accepted, sabotage could no longer be considered in either case. Catastrophic engine failure could also be eliminated, especially since inclusion of armour plating in the modifications. Medical evidence again pointed directly to break-up of the cabin owing to failure of a critical part of the aircraft structure from an unknown cause.

Action was taken to instigate a most thorough investigation involving the best brains in the country. The answer lay in the Comet 1 aircraft itself and in the pieces which lay at the bottom of the Mediterranean sea. The wreckage of Yoke Yoke lay in depths of water hopelessly beyond reach, so the search off Elba was intensified in the hope of finding as much of Yoke Peter as possible.

At Farnborough, work continued relentlessly under the supervision of a brilliant young Cambridge mathematician, Arnold Hall FRS. Director of RAE at only 36 years of age, he shouldered the burden of responsibility for the investigation and was later to be knighted for his work. While the wreckage of Yoke Peter already received was minutely examined, Comet 1 G-ALYU (Yoke Uncle) was withdrawn from BOAC's fleet and donated for destructive testing. The Comet engines retrieved from the depths off Elba were examined and given a clean bill of health. Catastrophic engine failure was satisfactorily deleted from the list of suspects.

Metal fatigue now became a more likely candidate, although it still seemed unbelievable to the scientists that it would prove significant so early in the lives of those Comets destroyed. If it was significant, failure of the wing was deemed most likely, but proof was required. Yoke Uncle was to be tested under simulated flight conditions with the wings and cabin being simultaneously subjected to repeated loads. By this method real time fatigue could be accelerated and the result analysed. Pressurising the cabin with air to the point of destruction would be equivalent to exploding a 500lb bomb, and was clearly unacceptable. Instead, a giant 'swimming pool' 112ft long, 20ft wide and 16ft high was constructed round the aircraft with the wings jutting out. The work took seven weeks to complete. Finally the wings were sealed with vulcanised fabric where they protruded from the sides, and the whole tank was filled with water. The effects of repeated cabin pressurisation could now be simulated by pumping water under pressure into the fuselage, holding the pressure steady at the cruise level value of 8.5lb/sq in differential pressure for a short time, and then reducing it to zero. Periodic tests were also to be made with the pressure increased to 11lb/sq in. Any failure of the fuselage would be safely contained by the water in the tank. Attached to the wings were hydraulic jacks which could be controlled electrically to simulate flight loads by manipulation. By such means the structural forces sustained throughout a three-hour flight could be simulated in five minutes. Soon a great number of tank 'flying' hours were accumulated in a comparatively short time to add to Yoke Uncle's actual flying hours before being grounded.

A round-the-clock surveillance of the tank was maintained in anticipation of a result as the weeks of testing continued. The staff at RAE regularly worked 80 hours a week over a period of six months and many of the scientists, including Arnold Hall, put in 100 hours a week. As suspected the first signs of metal fatigue were detected in the wings, and close inspection revealed small cracks beginning at the aft end of the bays which housed the undercarriage bogies in flight. After another 130 hours the cracks on the right side lengthened to over 8in and were considered serious enough to halt the tests. Repairs were conducted and the trials resumed. Had such a failure occurred in flight the result would have been disastrous. BOAC, however, had introduced at an early stage of the Comet operation a wing inspection programme which would undoubtedly have detected cracking of that magnitude before failure. The cause of the disasters seemed to lie in another area.

At the end of May, Yoke Peter's tail unit arrived at Farnborough to be placed like a giant jigsaw piece with the other sections. The port elevator spar which had suffered failure in the Calcutta crash was never found. It was believed from eye witness accounts that the tail unit had broken away early in the disintegration sequence, but a remarkable piece of detective work proved otherwise.

Assembly of the fragments aft of the wings showed a blue paint smear running from front rear along the outside of the left side of the fuselage and ending at the tail. The marks suggested an item being hurled backwards, scoring the skin of the aircraft and striking the left tailplane. Chemical analysis proved the paint to be the same as that used on aircraft seats! A small piece of carpet lodged in a gash in the tailplane supported the observation. Scratches were also found on top of the left wing running outwards from the fuselage towards the tip. Samples of paint specks imbedded in the scratches were taken and were shown to match that used on the fuselage. Both the wings recovered from the sea bed had outer wing sections detached, but here was absolute proof that the wings had been intact at the commencement of break-up. By the third week in June a piece of fuselage from the left side arrived at Farnborough. Its rough edges were shown to fit the scratches on the wing surfaces. All the evidence pointed to disintegration of the aircraft *following* catastrophic failure of the cabin.

Attention now focused on the cabin area, but the scientists, aware more than anyone of the excessive initial loading trials of the fuselage conducted by de Havilland during development, could not yet attribute a cause. Deterioration of the cabin structure, perhaps due to failure of the Redux bonding glue, was now considered a greater possibility than metal fatigue.

During a routine 'climb' of Yoke Uncle in the tank at the end of June 1954 the seemingly impossible happened. Pressure within the cabin suddenly dropped to zero. The tank was drained, and examination of the structure showed failure to have occurred at the corner of a cabin window. The scientists were not convinced.

Repairs were hastily conducted, the tank refilled and the tests continued. After a further series of routine 'flights' the pressure was again raised to the periodic level of 11lb/sq in. Suddenly, before reaching target, it again dropped to zero. Once more the tank was drained. The scientists were astounded. After only 1,800 test 'flights' in the tank, catastrophic failure of the cabin had occurred. Yoke Uncle had 'flown' the equivalent of only 9,000 hours, including flight time before being grounded. Its side was found to be ripped open with a break 8ft long and 3ft high along the left side of the fuselage above the wing. The failure point was once again the corner of a cabin window, demonstrating quite clearly that the loadings around such cutouts in the fuselage were far greater than expected. The cabin structure was much less resistant to fatigue loads than de Havilland's calculations had indicated.

But there was more. Apart from the cabin windows, there were another two cutouts positioned right on top of the Comet fuselage at the forward end of the centre section between the two wings. At these points were housed internally the two automatic direction finding (ADF) units and their aerials. Incoming signals from radio navigation beacons were received by these antennas and the aircraft bearing from the beacons was displayed on a compass for pilot guidance or for plotting position by the navigator. To permit radio reception the cutouts, known in the trade as ADF 'windows', were sealed by dark squares of glass fibre, a suitable non-conducting material. Close examination of the area round the ADF 'windows' revealed a number of tiny hair-line cracks emanating from rivet holes. Clear evidence of metal fatigue! Could this be the region of weakness which had spread throughout the cabin causing failure? Unfortunately none of the fuselage sections of Yoke Peter containing the ADF 'windows' or forward cabin windows had yet been recovered. No indication of fatigue could be found on the parts at Farnborough and proof, without these remaining pieces, was impossible.

While the tank tests were being conducted other experiments were being made which could shed light on the problem. Wooden models of the Comet 1 were constructed from segments and catapulted from the top of a tall building. From analysis of the disintegration information available the probable sequence of Yoke Peter's destruction was calculated. With disruption of the central section of the cabin, parts of the fuselage, nose and outer sections of the left wing had been torn off. The centre section with engines then separated and caught fire. The detached rear fuselage with tail fin struck the sea open end first, followed by the blazing centre section and wingtips. A line attachment distintegrated the model in the arranged sequence when it reached its extremity and pieces flew in all directions. The various trajectories of the model parts were used to calculate the probable scatter of Yoke Peter's wreckage and as a result the search off Elba was redirected to deeper waters. Trawling was the only means by which recovery could be achieved. At Farnborough the scientists toiled on in hope.

While all this work was going on, one of the grounded Comet 1s was permitted to do some flying for test purposes. Packed with electronic equipment and a few courageous scientists, the aircraft G-ANAV (Able Victor) flew a number of flights at high altitude and in severe manoeuvres. Although everything was done to minimise risk, the danger to those on board was readily apparent. The cabin of Able Victor was flown unpressurised although at the time there was no definite

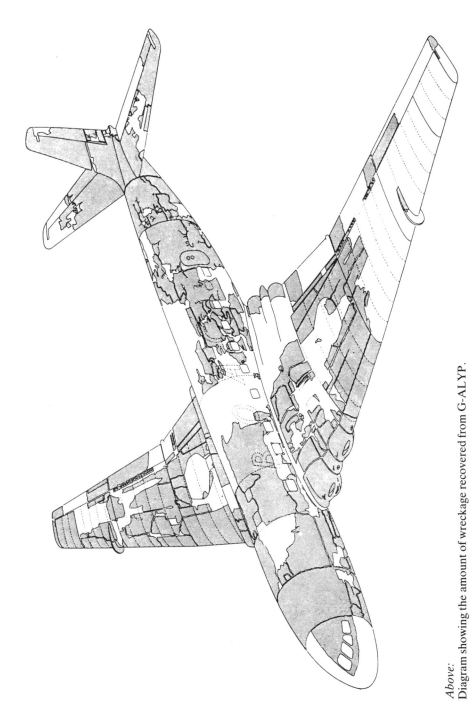

Above:
Diagram showing the amount of wreckage recovered from G-ALYP.

knowledge that danger lay in this region. Pressurised oxygen masks were available for the crew. As a precaution extra fire fighting equipment was placed aboard the Comet and a Canberra aircraft acted as chaperon during all flights to maintain observation. In all a total of about 100 flying hours were accumulated in three months and much useful information gained.

By late August the miracle had happened. Three more vital pieces of Yoke Peter had been found on the Mediterranean sea bed and returned to Farnborough. The parts included the all-important centre sections of the upper fuselage containing the ADF units and some forward cabin window cutouts. By now three-quarters of Yoke Peter's wreckage had been found. Examination of the area around the ADF 'windows' revealed metal fatigue cracks which clearly indicated weakening of the fuselage. It was also shown that rivets round the edges of two sides of the rear ADF 'window' cutout, and on one side of the forward cutout, had been placed too near the edge. Cracking had occurred at the rivets on both the skin and the reinforcing plate beneath. Some of this cracking was clearly a manufacturing fault, for holes were seen drilled at the ends of the cracks to prevent spreading. These manufacturing cracks in themselves, however, were not considered to be a danger, although they could have acted as a fatigue origin. Metal fatigue may have been accelerated as a result, but in any case the fatigue life of the cabin was proved to be low. Had the stress level at the cutouts not been so high there would have been no problem. Although some of the fatigue cracks may have run through manufacturing cracks, failure of the cabin was shown to have commenced at a crack of fatigue origin. All the breaks in the fuselage skin were noticed to run from the top cutouts housing the ADF units and were shown to be the root cause of the problem.

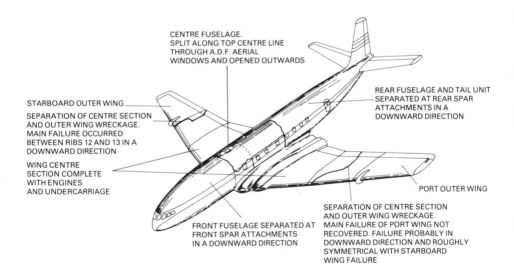

CENTRE FUSELAGE.
SPLIT ALONG TOP CENTRE LINE
THROUGH A.D.F. AERIAL
WINDOWS AND OPENED OUTWARDS

REAR FUSELAGE AND TAIL UNIT
SEPARATED AT REAR SPAR
ATTACHMENTS IN A
DOWNWARD DIRECTION

STARBOARD OUTER WING

SEPARATION OF CENTRE SECTION
AND OUTER WING WRECKAGE.
MAIN FAILURE OCCURRED
BETWEEN RIBS 12 AND 13 IN A
DOWNWARD DIRECTION

WING CENTRE
SECTION COMPLETE
WITH ENGINES
AND UNDERCARRIAGE

PORT OUTER WING

SEPARATION OF CENTRE SECTION
AND OUTER WING WRECKAGE.
MAIN FAILURE OF PORT WING NOT
RECOVERED. FAILURE PROBABLY IN
DOWNWARD DIRECTION AND ROUGHLY
SYMMETRICAL WITH STARBOARD
WING FAILURE

FRONT FUSELAGE SEPARATED AT
FRONT SPAR ATTACHMENTS
IN A DOWNWARD DIRECTION

Above:
Location and direction of main failures on G-ALYP.

The wreckage of Yoke Peter was compared with the failure to Yoke Uncle and the similarity in fatigue cracking confirmed. Subsequent calculations and strain gauge measurements demonstrated that the corners of the square ADF unit and cabin window cutouts were subjected under normal conditions to stress of as high as 70% of ultimate load. The figure was thought to be nearer 40% or 50%. On later Comets, fuselage strength was increased as a result of these findings and the cabin window cutouts designed oval in shape to eliminate stress concentrations.

The large amount of evidence amassed now pointed directly to the weak point of the Comet 1s, and the same fate could also be assumed to have befallen Yoke Yoke, although proof was still required. With almost no wreckage from the Naples crash, attention was now focused on the pathologists' reports. Both the British and Italian medical teams were of the opinion that the injuries sustained in the Elba and Naples cases were identical. Some dissension existed over certain aspects of the internal injuries received which the Italians attributed completely to explosive decompression and the British in some part to the force of striking water. Full-sized dummies were even dropped from aircraft to help calculate falling velocity and impact injuries.

During early development of pressurised cabins, tests had also been conducted on man-sized models to examine the effects of window blow-out. Loaded and fully clothed dummy passengers were placed in pressurised containers with a cabin window inserted. When the window was made to fail the sudden depressurisation resulted in the unfortunate dummy being sucked out through the window. Much information was available on such incidents although little was known of the effects of failure of an entire cabin. Could the results be reproduced experimentally?

At Farnborough a Perspex model Comet 1 was constructed to one-tenth scale, complete with tiny seats and miniature passengers. The Comet 1 model was placed in a chamber evacuated to an equivalent pressure at 40,000ft, while the cabin was pressurised inside to 8.5lb/sq in. The fuselage structure was then pierced at the top to simulate failure of an ADF 'window', and the resulting explosion photographed by a high-speed camera. Instantly a great gash ran from the bursting point along the top of the fuselage. In the cabin, during the first 1/100th of a second after initiation of the failure, little disturbance was noticed, but after 1/30th of a second there was movement forward of some of the seats. After 1/10th of a second the miniature passengers were seen to fly forward and upwards, striking the cabin roof with force. One tiny dummy passenger was seen to go straight through the gap which yawned. Seats were tossed about everywhere in the chaos.

The destruction of the model was conclusive proof of the pathologists' findings. The Italian report on the autopsies carried out after the Elba crash was shown to be a brilliant piece of deduction and its conclusions about the accidents, based on purely medical evidence, were remarkably accurate. The form of Yoke Yoke's destruction was now left in little doubt and could only have been the same as Yoke Peter's. After five months of relentless work by so many people, at a total cost of £2 million, the mystery of the Comet 1 crashes was satisfactorily solved.

The Court of Inquiry into the accidents opened on 19 October 1954 to examine the evidence. Over a five week period the inquiry sat for 22 days under the direction of Lord Cohen, and amassed 1,600 pages of shorthand transcript amounting to 800,000 words. Forty-four witnesses were called, 24 affidavits were read, and 145 exhibits were displayed. The RAE's researches into the cause of the accidents, amounting to a volume 4in thick, were presented at the summing up. In brief, the cabin was found to be prone to metal fatigue. It was weaker than expected and its working life shorter than realised. The wings and jet effluxes were also prone to fatigue.

Three months later Lord Cohen delivered his report:

'The accident was not due to the wrongful act or default or to the negligence of any party or any person in the employment of any party.'

'I am of the opinion that no blame can be attached to anyone for permitting the resumption of services.' (After the Elba crash.)

'The primary object of de Havillands was to lay the foundation for extensive tests which they regarded as the soundest basis for development of a project, rather than to arrive at a precise assessment of the stress distribution at the corners of the cabin windows. I do not think that they can justly be criticised for this approach to the problem. I am satisfied that in the then state of knowledge, de Havillands cannot be blamed for not making greater use of strain gauges than they actually did, or for believing that the static test they proposed to apply would, if successful, give the necessary assurances against the risk of fatigue during the life of the aircraft.'

The Farnborough investigation made aviation history. At a time when accident investigation and fatigue life estimation was just emerging as a precise science the RAE presented a demonstration of determined and dogged research leading to a solution which was admired throughout the world. The American magazine *Aviation Week* of February 1955 praised the result 'as the product of minute examination, probing analysis and deliberate deduction, held together through tortuous hours by British tenacity and national pride.'

At the end of the war de Havilland had been sufficiently courageous to venture into the unknown and to design and build the world's first jet transport aircraft. The two accidents had been the result of factors beyond the limit of contemporary knowledge, but the pioneer work undertaken by the company had paved the way for all future jet consideration. The subsequent analysis of metal fatigue provided a source of new information which benefited all manufacturers and which undoubtedly contributed to air safety. Experience gained from the demise of the Comet 1s established new loading and stress test standards throughout the world, and introduced improved methods of construction which quickly became the norm. Fatigue inspection became routine and soon developed into a science.

Having achieved a satisfactory result the case of the Comet crashes was closed, although a number of questions remained unanswered. The Comet 1 had been proved to be a weak aircraft, but the circumstances leading up to the break-ups were never examined. One major factor may be attributable to a crash, but since accidents owing to a single cause have been mostly eliminated, the result is normally the culmination of a sequence of events. During the test at Farnborough the cabin failed at the equivalent of 9,000 hours under the effects of repeated

Above:
BOAC Comet 1. *Via John Stroud*

Below:
G-ALYU in the test tank.

pressure application. At Calcutta, Elba and Naples the break-ups occurred at much lower flying hour levels: 1,649, 3,681, and 2,704 respectively. What had been the trigger in each case to bring about the catastrophe? The Calcutta crash was discarded completely from the Farnborough investigation but should perhaps have been examined in context. The aircraft broke up after failure of the left elevator spar in a thunderstorm, but it is now doubtful if the storm was of such severity that any aircraft would have been destroyed. It was simply too much for the weak Comet 1. At the time of the Elba crash strong winds blew over Western Europe. A jet stream with core winds in excess of 120kt ran due south from the UK over France, Spain and the Straits of Gibraltar, then turned at right angles in a large 'L' shape to blow eastbound across North Africa. There was severe turbulence in these areas. The West Italian coast lay on the periphery of the jet stream and, in spite of earlier reports, the air at altitude could at times have been subject to turbulence of sufficient severity to trigger cabin disintegration. At that time no other aircraft flew at these levels to verify reports. But what happened over Naples? Meteorological stations in the region simply noted no winds. Not, as first might be expected, no winds recorded, but simply no winds at all. Even at 30,000ft it was a flat calm. There is no doubt that the weak cabin broke up in a similar manner to the event over Elba, but what could have been the trigger under such exceptionally calm conditions? One event which was unknown at the time and which could have been a contributory factor is jet upset. In such circumstances a jet aircraft is grossly upset from its flight path by an instability factor which could be caused by trim or autopilot malfunction. In later years the Boeing 707 was shown to suffer from out-of-trim upsets. A small oscillation was known to affect adversely the aircraft automatic pilot trim which resulted in a progressive nose down force being applied. Eventually the effect would be sufficient to trip the autopilot and suddenly to plummet the aircraft into a dive. Exceptional measures were required to recover. If such an event had occurred suddenly to the Naples Comet 1 the somewhat crude control system could easily have seriously strained the aircraft. This can only be conjecture, of course, as there was no analysis of events which could have triggered the break-up, but it is a possibility. What actually happened can only be surmised as crash recorders were not available at the time.

But there is still more to the Comet crashes than has been previously discussed and many people, even today, discount the findings of the tests at Farnborough. Although there is little doubt that the Comet 1 was not a strong aircraft, whenever any attempt is made to analyse the break-ups, especially out of Rome, there are always several pieces missing from the jigsaw of events. It has since been established, for example, that in spite of earlier claims the failure of Yoke Uncle's fuselage in the tank was *not* at the same location as Yoke Peter's fuselage failure.

There are more than a few people who are convinced to this day that the earlier suspicions of sabotage on the Rome aircraft are well founded, and that either one or both of these accidents were caused by an explosive device. There have even been suggestions from responsible individuals that this was part of a plot to undermine the enormous advantage gained by the British in jet operations over the rest of the world. To begin with, Rome was not a secure airport. A small bomb could easily have been placed in a hold and could have detonated,

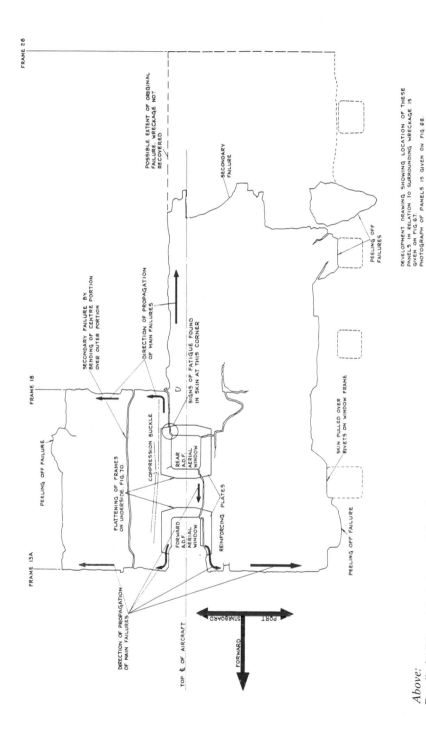

Above:
Detail of ADF aerial windows and the salient features of the disruption to G-ALYP's pressure cabin.

57

triggering the break-up of the weak cabin. It is certainly feasible that a blast, even in a hold, could have shaken the fuselage sufficiently to induce failure at a metal fatigued ADF window without leaving adequate evidence on passengers or aircraft. At that time there was insufficient medical or scientific knowledge to ascertain bomb damage without obtaining enough material from the seat of the explosion. Since the break-up occurred in the middle of the aircraft the bomb would need to have been planted centrally. Almost none of the central fuselage section of Yoke Peter was found and virtually nothing of Yoke Yoke. It wasn't until Comet G-ARCO was blown up near the Meditteranean Island of Rhodes in 1967 that sufficient knowledge was available to establish sabotage as the cause by examining material remote from the source of the blast.

The Comet 1 was permanently withdrawn and the Comet 2 production line halted, although RAF Transport Command agreed to take some of these after modifications. BOAC also accepted delivery of a modified Comet 2 with Rolls-Royce Avon RA29 jets for flight trials, engine operation, and crew training and experience, in preparation for the Comet 4. The Comet 3 had already flown by July 1954, but was used merely as a test vehicle for the proposed long range Comet 4. The Comet 3 incorporated the structural knowledge gained from lessons learned after the Comet 1 disasters, and it completed much of the certification testing for the planned new transatlantic Comet. In March 1955, BOAC placed firm orders for 20 Comet 4 aircraft to be delivered in 1958. Comet production was now assured. The new Comet was designed to carry a full load of 58 to 80 passengers at 500mph over a 2,870-mile range. One-stop flights from London to New York could now be achieved with services transitting Gander in Canada. Its structural strength was to be considerably greater than earlier models and a long working life was assured. In America, however, the sleeping giants were stirring and the Boeing 707 was already under development.

The Comet 4 first flew in April 1958 and was delivered to BOAC in the following months. Production of the Boeing 707 and Douglas DC-8 continued apace, with Pan Am finally accepting the Boeings. The race was now on between BOAC and Pan Am to operate the first transatlantic commercial jet service, with the competitors neck and neck. On 4 October 1958 BOAC's Comet 4 won by a nose, with one aircraft going into service on the London-New York route, and another simultaneously in the opposite direction, only 22 days before Pan Am's Boeing 707-120 operated from New York to Paris. Development of the Boeing 707 and DC-8 intercontinental versions, however, were already under way, with non-stop range between Europe and the US Eastern seaboard. The writing was on the wall for the Comets. Comet variants, the 4B and 4C, were built, but in the big jet sales boom which followed, none proved suitable contenders for long range operations.

In all, however, 113 Comets were completed before production ended in 1962. Of the 67 Comet 4s, 4Bs and 4Cs which entered airline service between 1958 and 1964, many provided excellent service lasting more than 25 years.

The Munich Air Disaster 1958

Just over a quarter of a century ago a great English soccer team won the affection of the entire British nation with their displays of football magic and skill. In 1957, Manchester United, with manager Matt Busby and his famous 'Busby Babes', were trying to emulate the previous season's efforts by once again attempting to scoop the honours at home and abroad. Not a man over thirty, and most of the talent home grown from the local area, the fresh young team attracted much support and interest from near and far.

By tradition the affairs of the great Manchester club were conducted in the manner of a different era. The first team players were treated like gods and were expected to behave and dress accordingly. The changing room was forbidden territory to all outside the elite, and even within the team a certain decorum was observed. The captain was expected to remain aloof from the other players and by custom lunched with the manager, not his team mates, before a match. Yet pay during the season was only £20 a week, with a £4 bonus for a win, and £17 a week in the summer months.

In the last season Manchester United, by way of reputation, had been invited as the first British club to participate in the expanding European Cup Championship. The team justified its renown by reaching the semi-finals where they were knocked out of the tough competition by a strong Real Madrid side, the eventual winners of the championship. In the same year they were also defeated by Aston Villa in the Football Association Cup Final, but managed to finish the season by achieving the League Championship, thus qualifying for another attempt at the European Cup.

Now in the 1957/58 season, Manchester United were battling successfully, if at times a little shakily, towards triumph in all three contests. In the European Cup, the first leg against the Czechoslovak champions Dukla brought victory at Old Trafford, the revered home ground, with a 3-0 win. On 4 December 1957, therefore, a confident Manchester United side flew to Prague to play the second leg of the tie on their opponents' own pitch. The result of the match and subsequent journey home both proved to be less than satisfactory. Manchester United suffered a 1-0 defeat at the hands of the Czechoslovaks in Prague but, with a win on aggregate, qualified for the quarter finals of the championship in which they were to meet the Yugoslav team Red Star. On arrival at Prague airport for the return, the flight was found to be indefinitely delayed because of fog at home, and it quickly became apparent that, with heavy bookings on other flights, there was a real risk of the team being unable to get back in time for their next league fixture. The party was eventually accommodated on two flights, with the team

Above:
'Elizabethan' class Airspeed
Ambassador.
Via John Stroud

Left:
Inside the Ambassador.
Via John Stroud

Below left:
Ambassador cockpit
detail. *BEA*

returning to Manchester via Amsterdam and the press via Zürich and Birmingham. It was a tired but relieved Manchester United entourage that finally arrived back in the familiar home territory of Old Trafford.

On 14 January 1958, the first leg of the quarter finals was won 2-1 at home against Red Star with the return match set for Wednesday 5 February in Belgrade. Welcome as the victory might be, the goal difference did not leave any room for complacency. To compound the situation Manchester United was scheduled to meet Wolves, the current Division One League leaders, in an important match on the Saturday afternoon following the Belgrade game. The tie could finally decide the eventual English League champions. Any disruption of flights to or from Yugoslavia similar to the return from Prague would place an unbearable strain on the already heavily taxed team. To alleviate some of the pressure it was decided that an aircraft would be chartered for the sole use of the complete Manchester United party including team, officials and press.

British European Airways (BEA) was approached and a forty-seven seat Airspeed Ambassador, named the 'Elizabethan' class by the airline, was duly acquired for the travellers. Thus, on Monday 3 February, the entire United ensemble boarded aircraft G-ALZU (known by the last two letters pronounced in the phonetic alphabet as Zulu Uniform), under the command of Captain James Thain for charter flight B-line 609 from Manchester to Belgrade. The route was arranged via Munich both ways for refuelling as a direct flight was out of range. At the airport the assistant team manager, Jimmy Murphy, was there to see them off. In his other capacity as manager of Wales he was required in Cardiff for a forthcoming international against Israel and was to remain behind.

On board, the party settled down for the journey that lay ahead while the captain checked the passenger manifest, reading the names like a 'Who's who' of British soccer greats. Roger Byrne, 28, the team captain and Manchester born; Geoff Bent; Jackie Blanchflower, brother of Irish international Danny; Bobby Charlton, brother of the equally famous Jackie; Eddie Colman, 21, a schoolboy discovery; Duncan Edwards, 23, once the youngest ever to represent England; Billy Foulkes; Harry Gregg, goalkeeper, recently bought for £23,000; Mark Jones, 24, another schoolboy discovery; Ken Morgans, 18, the youngest of the 'Babes' (once selected he was never dropped and during the current season had played 28 times for United with every game being won); David Pegg, 22, international; Tommy Taylor, most expensive player (four years previously transferred from Barnsley for £29,999); Dennis Viollet; Bill Whelan . . . an unprecedented list of British football talent of which the total complement was valued at the then astounding sum of £350,000.

With Matt Busby, himself a past player for the great rival Manchester City, as well as a former Scottish international, and now the much respected manager since his appointment after the war, were a dedicated threesome: Walter Crichmar, secretary, Bert Whalley, coach, and Tom Curry, trainer. Each had been with the club since the 1930s. The press was also well represented in the party with many reporters from local and national newspapers.

On the Sunday before departure, Captain James Thain, his co-pilot Kenneth Rayment and Radio Officer George (Bill) Rodgers had positioned the aircraft to Manchester from their BEA base in London in preparation for an early start.

Accompanying them were the cabin staff who would look after the party on the journey to Belgrade, Steward William Cable, and Stewardesses Margaret Billis and Rosemary Cheverton. In the cockpit, however, the flight crew were not the usual complement, as the co-pilot, Ken Rayment, was a full captain in his own right and slightly senior and more experienced than Thain. The co-pilot position would normally be filled by a first officer, but the two pilots were friends and had made a special request to fly together. The interest in common, however, was not football, but chickens! James Thain was 36, relatively young by today's standards, but at that time a senior captain and also chairman of the Elizabethan panel in the British Airline Pilots' Association (BALPA). He, like most pilots, had time between flights and some money to spare, and had set up a small poultry farm near his home in Berkshire which he ran with local labour. Captain Rayment had similar designs on starting in the poultry business and much talk had ensued between the friends. The original roster for the charter flight showed First Officer Hughes as the co-pilot but by request and a helpful administration office the crew list was changed to permit the two captains to fly together, thus allowing time before and after the match to continue the poultry farming discussions. Captain Thain was, of course, still the 'pilot in command' of the flight and was so documented, irrespective of the fact that a more senior and experienced captain, acting as co-pilot, sat on his right-hand side. The situation on the flight deck was somewhat unusual therefore, although not entirely uncommon. These circumstances were to have far reaching consequences for Captain Thain in events which lay ahead.

The trip outbound to Belgrade was not uneventful. The transit through Munich, relatively free from the effects of weather despite the time of year, was accomplished with little difficulty, but Belgrade was a different matter. The city was in the grip of winter with low cloud, poor visibility and snow obscuring the runway. Minimum limits of cloud base height and visibility are set at airports by each airline depending on certain items such as local terrain and availability of landing aids. When the weather deteriorates below these limits an approach to land is not permitted and if improvements are not forthcoming a diversion to another airport is required. The conditions at Belgrade were precisely on limits and demanded some skilful flying in the 'let-down' pattern, as the aircraft had to break cloud at just the right point and height for a final approach and landing to be attempted. At that time BEA employed the 'monitored approach' procedure, a system which is still used by some airlines to this day. In this case the pilot who is operating the sector and who will land the aircraft does not handle the controls during the approach but monitors the other pilot flying the procedure on instruments while frequently peering into the cloud for sight of the runway. With the landing strip in view the operating pilot then takes control and lands the aircraft. Such a system avoids the sometimes difficult transition from instruments to visual flying as the aircraft breaks cloud when one pilot flies the entire approach and landing. During the approach to Belgrade Captain Thain, as operating pilot, had relinquished the flying to his co-pilot, and only a masterly piece of instrument flying by Rayment allowed Thain successfully to take control at cloud break and to land the aircraft; a fine demonstration of Ken Rayment's ability and skill. In the prevailing conditions the station engineer was not expecting an imminent arrival, and only realised that Zulu Uniform, the Manchester charter flight, had landed when it taxied on to the parking apron.

By Wednesday 5 February 1958 the weather had markedly improved with clear skies and sunshine. The pitch at 'The Army' football ground was cleared of snow, except for a few specks, and a large and enthusiastic crowd gathered in the packed stadium to watch the match. Football fans throughout Europe waited in anticipation of the result. The game proved to be hard, tough and sometimes dirty, but by half-time Manchester United were leading 3-0. Throughout the second half Red Star rigorously chipped away at United's lead with one goal following another until the final whistle blew with the score level at three all. A tired but mightily relieved Manchester United team were once again, on goal average, through to the semi-finals of the European Cup for the second year running. After the match a cocktail party for Manchester United followed at the British Embassy with much larking and clowning amongst the young and jubilant team. At home British fans rejoiced at the victory.

The next morning, Thursday 6 February 1958, the still high-spirited team assembled with others of the party at Belgrade airport for the return flight to Manchester, via Munich. In addition to the group, an extra five people travelling to Britain, including the wife and baby daughter of the Yugoslav air attaché in London, were offered seats on the flight, bringing the total boarding the Elizabethan to 38. On board the 47-seat aircraft, therefore, there was sufficient space for the passengers to seat themselves comfortably, arranging positions in the cabin as they pleased.

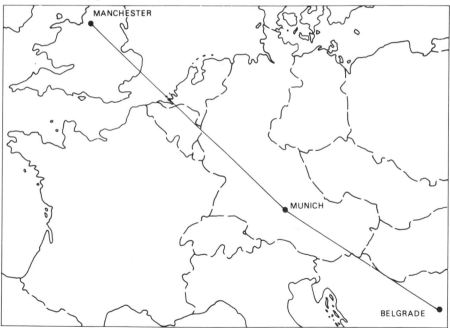

Above:
The intended route from Belgrade to Manchester.

On the flight deck similar seating arrangements were also being discussed. Normal practice amongst most flight crews is for the two pilots to share the flying equally and the Manchester United charter flight was to be no exception. Since Captain Thain flew the two legs of the outbound journey to Belgrade, his co-pilot, Rayment, was to fly Zulu Uniform back. That created one problem which was simply and amicably solved. Company regulations stated quite categorically that the pilot in command, in this case Thain, must always occupy the left-hand seat, although Civil Aviation Authority (CAA) regulations did not forbid changing positions. Rayment, although acting as co-pilot in the right-hand seat, normally flew the aircraft as captain from the left-hand seat. He was used to handling the aircraft with right hand on the throttles and left hand on the flying controls, and was more familiar with the pattern of movements required from the left-hand seat during take-off and landing. In the right-hand seat his actions would not flow naturally. Moreover, certain instruments placed on the right side could be observed properly only from the right-hand seat, and it would be Captain Thain's duty, now the non-handling pilot, to monitor these dials during the critical stages of the take-off and landing. The two captains followed what they thought was the most sensible and reasonable course of action in the circumstances and simply changed seats. It had been done before in similar situations and would cause little more than mild embarrassment if the breaking of regulations were brought to the attention of the company by something untoward occurring. It could not be known by either captain that only a short time hence something very untoward was about to happen which would have catastrophic consequences for both of them.

The departure from Belgrade was pleasant enough in intermittent sunshine, but the forecast for Munich was less than satisfactory, with low cloud, rain and snow. The descent into Munich through 18,000ft of cloud required switching on of the airframe anti-icing which permitted hot air heated to 60°C by fuel burners in the wings to be vented under pressure via ducting at the leading edges of the wings, tail plane and fin. The aircraft broke cloud at 500ft above the ground to a grey, bleak and overcast day, with the runway only just visible where previous aircraft had left black streak marks in the slush and snow. The touch down was normal, but on roll out great plumes of wet slush sprayed upwards from both sides of the nose wheel. The drizzle which had been falling for some time had now turned to snow, although the temperature remained above freezing, and Zulu Uniform had to taxi across the apron through a watery paste of slush and snow to the parking bay by the terminal building. The time was just after 13.15hrs GMT (14.15hrs local).

As the flight crew walked through the wet snow on the tarmac to the airport Meteorological Office for the onward weather briefing, refuelling had already begun. Bill Black, the station engineer, supervised the refuelling of the two 500gal tanks (one in each wing) but, as the only licensed Elizabethan engineer on duty, he could not, for technical reasons, permit use of the underwing pressurised fuelling system. The local ground staff had to resort to the use of the emergency fillers on top of the wings. Although snowing, the wings were warm from the anti-icing applied on descent and, with the temperature above freezing, any flakes settling on the upper wing surfaces simply melted and ran from the trailing edge.

PLAN OF MUNICH AIRPORT
February 1958

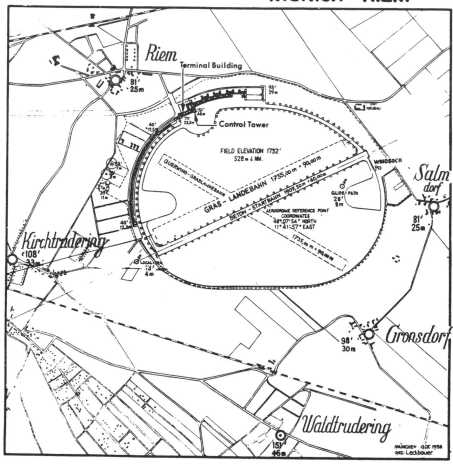

MUNICH–RIEM

Riem

Terminal Building

81'
25 m

95'
29 m

NEUBAU

15'
18 m

73'
22.5 m

Control Tower

40'
+12.5

FIELD ELEVATION 1732'
528 m ü. NN.

QUERWIND - GRASLANDEBAHN

GRAS - LANDEBAHN 1755,00 m × 90,00 m

BETON - STARTBAHN 1907,50 m × 60,00 m

AERODROME REFERENCE POINT
COORDINATES
48°07'54" NORTH
11°41'57" EAST

1735,00 m × 90,00 m

WINDSOCK

GLIDE-PATH
26'
8 m

Salm dorf

81'
25 m

11 m

40'
12.5 m

Kirchtrudering
108'
33 m

LOCALIZER
13'
4 m

98'
30 m

Gronsdorf

Waldtrudering
151'
46 m

MÜNCHEN, den 1958
gez. Lechbauer

1 NAUTICAL MILE

1 KILOMETER

SCALE 1:20000

LOCAL MAGN.VAR. JAN 1953   2°54' WEST

○   OBSTRUCTION, ILLUMINATED

○   OBSTRUCTION, NOT ILLUMINATED

·   TAXIWAY - LIGHTS OUTSIDE - RIM = BLUE

·   TAXIWAY - LIGHTS INSIDE-RIM - YELLOW

▬   OBSTRUCTION LIGHTS ON AIRPORT-BUILDINGS

ᵠ   WINDSOCK WITH OBSTRUCTION LIGHT

Above:
Munich airport plan.

65

The German refueller, therefore, had little problem maintaining a foot grip as he walked over the wing, which was just as well, for he soon became an easy target for the Manchester party's snowballs as they disembarked for refreshments in the terminal building.

After completion of departure details the flight crew returned to the aircraft and before boarding stopped to inspect the wings. As a precaution de-icing fluid could be sprayed on upper surfaces, or any snow accumulation brushed clear, but neither seemed necessary under the circumstances. For those unfamiliar with aerodynamics it is important to point out here that aircraft lift is obtained from an airflow speeding up as it flows the long way round over the upper surfaces of the wing, causing a reduction in pressure. Any ice or snow accretion on upper surfaces deteriorates lift production by spoiling the airflow and all upper surfaces must be maintained free of snow or ice. On inspection of Zulu Uniform's wings a thin film was just visible, although already thawing, and water could be seen dripping from the trailing edge of the wings. The two captains conferred, and their decision not to proceed with de-icing passed to the station engineer. Captains of other aircraft, however, seemed to require de-icing to be undertaken, more from a 'matter of routine' attitude than anything else, as it strictly wasn't required.

At 14.20hrs GMT (15.20hrs local), just over one hour after arrival, with the lively band of passengers all aboard, Zulu Uniform called Munich control tower to apply for taxi clearance for the final leg of the journey to Manchester, and home.

ZU R/T: 'Munich Tower, B-line 609 Zulu Uniform, taxi clearance, please.'

Munich Tower R/T: '609 Zulu Uniform, München Tower, wind two nine zero, eight knots. Cleared to runway two five. [The magnetic direction of the runway rounded up to the nearest 10° with the last digit omitted; eg the actual runway magnetic direction of 246° (degrees) becomes 250° becomes 25.] QNH [mean sea level pressure setting in millibars] one zero zero four, time one nine and a quarter, over.'

ZU R/T: 'Roger, thank you.'

The runway was known to be covered by about 2cm of slush which, at that time, was not considered excessive, especially since a path seemed to have been trodden through much of the surface used for take-off and landing by previous departing and arriving aircraft. Two minutes later B-line 609 approached the threshold and was cleared on to the runway.

Munich Tower R/T: 'B-line 609, cleared to line up and hold, and here is your clearance, over.'

ZU R/T: '609, understand. I got to line up and hold and I am ready for the clearance.'

Munich Tower R/T: 'Munich control clears B-line 609 Zulu Uniform to the Manchester airport via amber airway one zero, green airway one, amber two, amber one and route as filed. Maintain one seven thousand feet. Right turn after take-off. Climb on the south course inbound to the Freising range, over.'

Zulu Uniform repeated the clearance with the incorrect statement that a southbound heading inbound to the Freising radio range was required and a request was made for confirmation.

Munich Tower R/T: '609, that is not correct. Climb on the south course of the Freising range inbound to the range, over.'

ZU R/T: 'Right turn out, the south course inbound to the Freising radio range, over.'

Munich Tower R/T: 'Ah, that is correct.'

The aircraft lined up on the runway, and, with brakes confirmed set, power was opened up for a final engine run in order to check performance before departure.

Munich Tower R/T: 'B-line 609, how long will it take for the engine run, over.'

ZU R/T: 'Half a minute.'

Munich Tower: 'Ah, Roger. Your clearance expires at three one. Time now three zero.'

With the engine run hastily concluded all drills were now completed in preparation for departure.

ZU R/T: 'Ah, Munich, 609 Zulu Uniform — I am ready for take-off.'

Munich Tower: '609 Zulu Uniform, wind two nine zero, one zero knots, cleared for take-off, right turn out.'

ZU R/T: 'Thank you.'

As the aircraft moved off Radio Officer Bill Rodgers made a final call indicating that the run had begun.

ZU R/T: 'Rolling.'

Captain Rayment opened up the throttles and moved both levers slowly forward with his right hand while Thain followed through with his left. With the throttles against the stops Thain rapped Rayment's hand as an indication he should relinquish control, then made some final adjustments to the setting. As procedure demanded Rayment requested 'Check full power', while his right hand remained poised ready to close the throttles quickly in the event of a full stop being required. Thain confirmed 'Full power set', and then a little later 'Temperatures and pressures OK, warning lights out'. The aircraft continued to accelerate with little indication of the retarding effect of the slush while Thain called out the increasing speeds. Suddenly an uneven tone in the engines was detected which spelled alarm. A quick instrument scan was in order, but just as Thain's eyes caught the fluctuating pressure gauges Rayment called out 'Abandon take-off'. His right hand slammed closed the throttles, catching Thain's fingers before he could move to hold forward the control column as Rayment applied full brakes. Zulu Uniform decelerated rapidly and slithered to a near halt some way down the runway.

ZU R/T: '609, we are abandoning the take-off.'

Munich Tower: 'Tower, say again please.'

ZU R/T: 'We are abandoning the take-off. May we backtrack, over.'

Munich Tower: '609, cleared to backtrack.'

ZU R/T: 'Thank you.'

The problem was the familiar one of 'boost surging'. Neither captain had been happy with the uneven engine note and fluctuating pressure and both were in agreement over the termination of the take-off. As Thain blew on his skinned knuckles Rayment apologised for the injury, having little need to explain that the moment had not called for delay.

'Boost surging' had been experienced by both pilots in the past as the Elizabethan had been somewhat prone to the problem when first introduced into service, and Rayment himself had had similar experiences at Munich. The cause was due to an over-rich fuel mixture upsetting the fuel distribution and resulting in some cylinders operating in a rich condition. The problem was therefore more acute where the air was thin at high altitude airports such as Munich which lay only 70 miles north of the Alps at an elevation of more than 1,700ft. Any power reduction resulting from 'boost surging' was stated to be only marginal, but it was hardly surprising that pilots were somewhat sceptical of the theory and rather wary of the uncomfortable condition. One answer to boost surging was to open the throttles slowly, so it was decided to try the take-off once more, applying this approach.

As Zulu Uniform started to taxi back down the runway for a second attempt a transmission from the tower interrupted the discussion.

Munich Tower: '609, for your information, we have a car on the runway which will be removed any second.'

ZU R/T: 'Have a car on the runway?'

Munich Tower: 'Disregard. The car is leaving the runway now — it's at the other end.'

Meanwhile on Zulu Uniform's flight deck the deliberation continued. It was agreed the best procedure to adopt was firstly to open up the throttles a little on the brakes before commencing the roll, and then gradually to ease them up to full take-off power during the run. Thain was to keep a careful eye on the boost pressure and rpm and to call out if there was anything he didn't like. In the cabin there was little consternation over the quick stop on the runway. After all, the weather was not particularly good and the abandoned take-off could have been the result of any number of factors. Now the problem seemed to be solved and they would soon be on their way.

Approaching the threshold of runway 25 clearance for departure was requested once more.

ZU R/T: 'Munich from 609 Zulu Uniform. When we get to the end of the runway we should like to take-off again. Is the clearance still valid?'

Munich Tower: '609, your clearance is still valid. However, maintain five thousand feet until further advised. The wind three zero zero, eight knots, cleared for take-off.'

ZU R/T: 'Thankyou. Understand, maintain five thousand feet until further advised.'

With departure checks re-completed the aircraft was lined up with the runway and the throttles opened up on the brakes to about 28in of boost. The power was confirmed set by both pilots and the brakes released. Zulu Uniform began its second take-off run.

ZU R/T: 'Rolling.'

Rayment slowly opened up the throttles against the stops and requested a 'full power check'. The aircraft was now accelerating down the runway. Thain's eyes scanned the instruments and confirmed the power setting. 'Full power ch... .' His response was interrupted by a rising pressure gauge. The starboard engine was steady at 57.5in boost pressure but the port indicator continued through 60in and went off the clock. 'Abandon take-off.' This time it was Thain who called out. The throttles were quickly closed with full braking applied, and the aircraft once again slowed to taxi speed half way down the runway.

ZU R/T: 'Munich, 609 Zulu Uniform, we are abandoning this take-off as well.'

Munich Tower: 'Ah, Roger, cleared to backtrack.'

The procedure of gently opening the throttles had worked for one engine but not the other and it was necessary to take further stock of the situation.

ZU R/T: 'We would like to be cleared back to the tarmac, over.'

Munich Tower: 'Roger, cleared to the ramp.'

ZU R/T: 'Understand, cleared to backtrack and cleared to the tarmac. Affirmative.'

Munich Tower: 'You are cleared to backtrack or cleared to taxi via crosswing — as desired.'

ZU R/T: 'Roger.'

Munich Tower: 'B-line 609, do you wish to make another take-off right now or do you wish to wait at the ramp? over.'

ZU R/T: '609 Zulu Uniform, I am returning to the tarmac. I am returning to the tarmac — to the terminal, over.'

Munich Tower: 'Roger, that is understood.'

The taxi from the runway back to the parking bay was achieved not without difficulty. While Rayment spoke to the passengers Thain took control and had some trouble in following the lines of the taxiway with the fresh fall of snow. In the cabin the Manchester United party listened attentively to the announcement explaining that a technical fault had prevented departure and it was now necessary to return to the apron for further checks. As the aircraft parked once again at the bay the tower was called with the message '609 Zulu Uniform at the ramp'.

It was a less than jovial party that disembarked for a second time at Munich just 20min after they had first departed. There was little doubt they were in for a long wait. Coffee was soon ordered for the group, and any feeling of apprehension quickly dispelled by the usual round of larking and joking. On the flight deck Bill Black, the station engineer, joined the two pilots for further consideration of the boost problem. The recommended procedures had been followed with some success, and the only other answer was to retune the engines which would mean night-stopping the aircraft. All agreed that this seemed a bit drastic. Since one engine was operating satisfactorily it was felt that if the throttles were opened up even more slowly there would be a good chance of eliminating the problem. If any fluctuation was still encountered it would be only marginal. It would, of course, mean take-off speed being achieved farther down the runway, but since there was plenty of distance available it wasn't going to pose a problem. The two captains concurred and made the decision to give it another try.

There was still, however, the question of de-icing being required and the pilots talked over the situation with the ground engineer. The snow continued to fall, of course, but any accumulation on the wings would have been adequately blown free with the two attempted take-offs. Although the total upper wing surface could not be fully observed from the flight deck it could be seen that the thin film over the wing noticed earlier had dispersed and the odd fleck lying on the upper surfaces could hardly cause problems before the next take-off. The same, however, could not be said of the runway, although in line with contemporary knowledge little or no concern was expressed about the surface state. A West German government minister was due later that day, and the airport vehicle, which had earlier briefly interrupted Zulu Uniform's backtrack, now conducted an official inspection of the runway in anticipation of his arrival. About one centimetre of slush was reported as the minimum depth visible throughout the entire length, although pilots' observations indicated that many areas were much deeper. Tracks were seen to be beaten over the first two-thirds by previous aircraft movement but the last third of the runway was covered in an unbroken layer of wet slush. The survey was rapidly carried out with only cursory measurements being taken at a few points, yet the situation was considered to be satisfactory. On completion of the inspection a Convair took off, spraying up great waves of slushy snow.

It seems inconceivable, admittedly with hindsight, that no attempt whatsoever was made to sweep at least some of the runway clear, but at the time there was little or no appreciation of the dangers of operating in such circumstances. The total lack of knowledge then becomes self-evident when it is considered that the present procedures for operating even something as big as a Boeing 747 from a contaminated runway are very restrictive. Regulations permit a maximum depth at take-off of only 0.5in slush (13mm) or 1.5in dry snow (38mm). Nowadays motorists driving in winter conditions also understand the problems associated with slush encounter; forward speed is immediately retarded and steering capability is markedly reduced. It is not difficult to imagine the effect of several centimetres of slush on aircraft acceleration. Any effort to clear some of the runway, even with a gang of men with shovels, would have been better than nothing. And yet that's exactly what was done — nothing. Within the bounds of understanding at that time a few centimetres of slush were thought to be satisfactory.

Flight 609 Zulu Uniform was now attracting some attention after the two abandoned take-offs, and a number of people, including the airport director in his first floor office awaiting the arrival of the government dignitory and a few airport trainees on the level above, stopped to watch the departure preparations. Someone even took a photograph. Zulu Uniform was becoming the centre of attraction and was not to be allowed to leave unobserved. With the decision now made to try again the passengers waiting in the airport lounge were duly summoned less than 10min after disembarking. Amid some surprise that departure was called so soon, and with coffee half drunk, the Manchester party somewhat reluctantly boarded the aircraft for the second time in anticipation of a third attempt at take-off. There was little fooling now except for a few nervous wisecracks and the atmosphere was decidedly tense. Matt Busby was overheard in conversation saying that 'if it wasn't so important for the lads to get back to Manchester for a good night's sleep before Saturday we could return tomorrow morning'. As the passengers settled aboard just before 15.00hrs GMT (16.00hrs local) the door was closed and Captain Rayment lifted the handset to speak a few words. The passengers were informed that the technical fault had been resolved and they would shortly be on their way. The flight time was to be just under three hours, and with a one hour time change an estimated arrival in Manchester was given as 18.00hrs local.

ZU R/T: '609 Zulu Uniform, Munich Tower. Would you have my clearance renewed, I am about ready to start.'

Munich Tower: 'München Tower, Roger.' Then a few seconds later. 'B-line 609, cleared to start engines. Your flight plan has been delivered to ATC [Air Traffic Control].'

ZU R/T: 'Munich Tower, B-line 609 Zulu Uniform, I am ready to taxi.'

Munich Tower: '609 Zulu Uniform, München Tower. Wind two nine zero, eight knots. Cleared to runway two five. QNH one zero zero four. Time five six and three quarters, over.'

ZU R/T: 'Thank you.'

The aircraft moved off across the tarmac for the three-minute taxi to the runway as the final departure checks were being completed. Approaching the threshold the tower was contacted for permission to enter.

ZU R/T: '609 Zulu Uniform, are we cleared to line up?'

Munich Tower: 'B-line 609 Zulu Uniform, cleared to line up and hold, and here is your clearance — München control clears B-line 609 to Manchester airport via route as filed. Maintain one seven thousand feet. Right turn after take-off, climb on south course inbound Freising range and maintain four thousand feet until further advised, over.'

The clearance was read back, this time without error, and the aircraft positioned on the runway in preparation for the engine run check.

Munich Tower: 'B-line 609, what is your rate of climb? over.'

ZU R/T: 'Six hundred feet a minute.'

At the end of the engine run a further transmission from tower interrupted the departure preparations.

Munich Tower: 'B-line 609, the clearance void if not airborne by zero four. Time now zero two.'

ZU R/T: 'Roger, understand, valid till zero four.'

A final few words were exchanged between the captains and it was agreed that Thain should keep his eyes glued to the engine instruments and adjust the throttles himself if any surging occurred.

ZU R/T: 'Ah, Munich, 609 Zulu Uniform is ready for take-off.'

Munich Tower: '609, the wind three zero zero, one zero knots, cleared for take-off.'

Commencing take-off the throttles, as before, were slowly moved up to 28in of boost and the brakes released.

ZU R/T: 'Rolling.'

Gingerly, Rayment advanced the throttles inch by inch, followed through by Thain who, as duty demanded, kept his attention fixed inside the flight deck monitoring the engines. Acceleration was slower than normal as the throttles were ever so gently opened to full power. Even in the cabin the slow engine response was noticed. As the aircraft gathered speed thick showers of slush were thrown up from the undercarriage and could be seen by the passengers on either side. Thain followed carefully through with his left hand as Rayment pushed the throttles against the stops with his right. Once again Thain tapped the back of

Rayment's hand and took control of the throttles. At Rayment's request to 'Check full power' Thain, with eyes fixed on the gauges, replied 'full power set, temperatures and pressures OK.'

Now established in the take-off run with all well and full power set, acceleration continued, albeit at a somewhat slower pace, with Thain calling the speeds . . . 60kt, 70kt . . . At around 85kt Thain suddenly called out that the port engine was surging slightly. While Thain dealt with the surging, Rayment transferred from nosewheel steering to rudder control as the rudder became fully effective with gathering speed. Rayment now pulled gently back on the control column to raise the nose about 4° above level and lift the nosewheel free from the slush in preparation for take-off. The simple manoeuvre was achieved not without difficulty. Meanwhile Thain eased the left throttle rearwards until the surging ceased with the boost pressure reading 54in. Slowly but steadily the port throttle was pushed forward again until fully open. Both boost pressure indicators now showed 57.5in with no evidence of surging. Thain confirmed once more that full power was set with temperature and pressures OK then returned his attention to the airspeed indicator. The speed showed 105kt. Acceleration was continuing slowly but to Thain, with attention still in the cockpit, it did not seem abnormal. Thain now called 110kt as he watched the airspeed needle move more sluggishly round the dial, flickering as it did so as if striving to grasp every knot. At 117kt Thain called V1, the decision speed, the point after which there was insufficient runway to stop. Zulu Uniform was committed to take-off. Rayment now made a slight adjustment to aircraft trim to ease the strain of holding the nose wheel off the ground while Thain's eyes remained fixed on the airspeed indicator. The next speed call would be at 119kt, V2 (the minimum safe speed required in the air following an engine failure at V1, the worst possible moment for such an incident). After V2, Rayment would be free to pull further back on the control column to fly off the ground and the aircraft would then be at a safe flying speed even if loss of power occurred one one engine.

Zulu Uniform was now well down the runway and was approaching the area of even slush untrodden by previous aircraft movements. Acceleration was negligible and the airspeed indicator still hesitated around 117kt. Suddenly there was a marked drop of about 4-5kt and for the first time Thain had the distinct feeling of lack of acceleration. The pointer then dropped and faltered at about 105kt. The end of the runway was now approaching with insufficient speed for flight and inadequate distance to stop! As the pilots confronted their dilemma Rodgers, the radio officer, continued with his duties by transmitting a last message to the tower.

ZU R/T: 'Munich, from B-line Zulu Unif. . . . '

The aircraft left the paved surface and ploughed through the snow towards the boundary fence. Rayment shouted 'Christ, we won't make it!' Thain looked up for the first time to see the dramatic scene before him as Zulu Uniform, already 200yd from the end of the runway, tore through the fence and on across a small road on the other side. Ahead, immediately within their path, lay a house and a tree. With his left hand Thain banged the already fully open throttles while Rayment tried in vain to pull Zulu Uniform off the ground. Rayment called for the undercarriage to be retracted in a desperate attempt to do something to become airborne. Thain quickly selected the gear up and the aircraft's movements became smooth as if flying through the air. Thain gripped the instrument coaming with both hands and the two pilots watched helplessly as the aircraft began to turn slowly right on a path between the house and a tree from which there was no escape. The controllers in the tower, blind to the drama in the falling snow, heard only Rodgers's last attempt at a message followed by a sound which, in the descriptive words of the official report, 'starts with a howling-whistling noise and ends with a loud background noise after the message was broken off.' From Zulu Uniform's cockpit it could be seen that a collision was unavoidable and Thain ducked his head behind the coaming as the aircraft struck the house on the opposite side of the small road. The impact tore the left wing off outboard of the engine and ripped off part of the tail unit, setting the house on fire. The stricken aircraft spun out of control, its momentum carrying it beyond the house and into the tree which struck the left side of the flight deck tearing open the cockpit. One of the undercarriage wheels detached and spun off towards a vehicle on the road while the broken craft continued to slither on through the snow. A hundred yards past the house the right fuselage aft of the wings struck a wooden garage containing a truck, severing the complete tail section. The truck's petrol tank exploded enveloping the shed in flames. The forward remaining section of Zulu Uniform ploughed on for a further 70yd before coming to a halt while the port engine, which had now detached, slid forward on its own a few yards ahead. The series of impacts enveloped the plane's occupants in a cacophony of breaking, tearing and crashing noises while they were violently shaken and spun in the burning aircraft. Then all at once, stillness and complete quietness. It was as if the din and uproar of smashing metal had been instantly switched off. No one spoke or shouted, no one cried out. Not a sound stirred in the eerie silence. It was quite uncanny.

Rayment, sitting on the damaged side of the flight deck, was badly injured, but Thain, unhurt, quickly came to his senses and gave the order to evacuate. Rodgers threw open the battery master switch and pulled a number of circuit breakers to shut off dangerous electrical circuits which could prove a further fire hazard, then squeezed through the emergency window of the galley door. The exit had been dislodged with the impact although escape was still possible, but the door leading to the passenger cabin was jammed solid behind a heap of baggage. Thain now leapt up to follow Rodgers but Rayment was struggling in his seat. His foot was caught and he made it known that he was stuck and unable to get out. Thain urged him to get out quickly but he couldn't move so Rayment suggested Thain go on ahead. Thain then crawled through the same emergency window exit used by Rodgers and quickly tried to assess the situation. There were a number of fires around the wreckage: flames could be seen at the stub of the left wing and also below the right wing with its 500gal fuel tank still intact. There was a great risk of a violent explosion. Standing amongst the wreckage were the two stewardesses who had exited by a crew door, but many of the passengers, stunned and dazed, still sat in their seats. Thain shouted to the girls to get away from the aircraft. The *Daily Mail* photographer, Peter Howard, fumbling about in a bewildered state, stumbled upon a gap in the fuselage and simply crawled out on his hands and knees, closely followed by his assistant, Ted Elyard. Harry Gregg, the goalkeeper, intact apart from the loss of his shoes and with little more than a bloody nose, managed somehow to struggle unshod from the wreckage. Meanwhile Thain and Rodgers, ignoring the danger of explosion, clambered back into the broken machine to grab the two flight deck hand-held fire extinguishers, pausing to assure Rayment that when the fires were out they would be back to help. As Thain discharged the extinguisher at the fires by the broken wing, he noticed through a window Billy Foulkes, the right back, still sitting in his seat, stunned by the impact. He shouted to him to get out quickly as the aircraft was liable to go up at any second. Foulkes could feel himself trapped in the seat and was beginning to panic, but suddenly remembered he was still restrained by the seat belt. Quickly undoing it he checked his legs for injury then leapt through a gap which had opened up before him. Once free he took Captain Thain's advice and just ran and ran, covering about 200yd before daring to stop and look round.

Soon the hand-held extinguishers, useless against such a conflagration, were spent and discarded, and a thick column of black smoke rose from the flames into the grey sky. Undeterred by the blaze and the imminent danger, a number of able bodied survivors — Rodgers, the two stewardesses Bellis and Cheverton, Elyard, Gregg and Howard, joined by the returned Foulkes — re-entered the stricken aircraft to help those trapped inside. Thain returned to aid Rayment, still pinned in the cockpit. In the broken cabin Matt Busby was found seriously injured near the rear of the wreckage propped up on one elbow, clutching his ribs, and a coat was rolled beneath him for support. Forward of him Bobby Charlton, slumped motionless in his seat, was still fastened by the belt, while Dennis Viollet, beside him in the same row by the window, sat similarly inert. Both appeared beyond help. As the rescuers approached, Charlton stirred as if shaking off sleep, sat upright to undo his safety strap, then simply stood up and walked towards them. Dennis Viollet followed suit! On the left, Jackie Blanchflower was alive but nursing a badly cut arm which was quickly dressed with a tie. It seemed, in spite of the wreckage, that the casualty figures, at least for those in the forward section of the aircraft, were light. Little did those survivors know that of the 44 souls on board, comprising 38 passengers and six crew, 20 had already lost their lives. Eleven members of the Manchester party had perished, including seven players, the coach, trainer, secretary and one director. At just after 15.00hrs GMT on that bleak winter's day in Munich, in an area not much bigger than a football field, the hopes, aspirations and dreams of a great soccer team lay scattered in ruins. The ranks of top journalists were equally devastated with seven leading soccer correspondents among the dead and one very badly burned. The steward and one Yugoslav passenger had also been killed.

Small fires continued to burn throughout the aircraft but the feared explosion had not occurred and rescue work was able to continue. On the smashed flight deck Thain's attempt to free the still trapped Rayment had proved unsuccessful and it became obvious that cutting equipment was needed to obtain his release. Soon first aid teams and ambulances appeared on the scene, followed shortly by the fire service. The flames around the wreckage were quickly brought under control, and those seriously injured were hastily put in ambulances for the desperate rush to hospital. Thain, still deeply concerned about Rayment's predicament, borrowed a fireman's axe and tried to hack a way through the

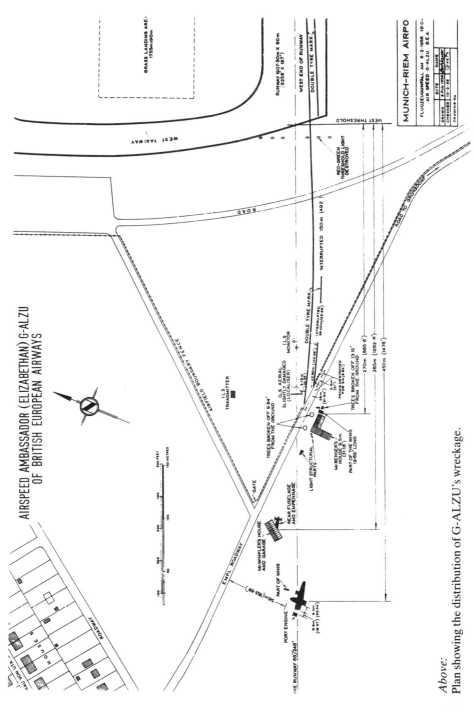

Above:
Plan showing the distribution of G-ALZU's wreckage.

cockpit sidewall, but to no avail. Eventually rescuers persuaded him of the need for a medical check up, and so, with thumbs up to Rayment, he was led to a waiting vehicle and driven from the scene of the crash. Rayment smiled bravely in reply to Thain's signal and waited calmly to be extricated. Later, rescue workers climbed onto the cockpit roof via the starboard wing and managed from above to free the badly injured Rayment from the tangled metal. Soon he was rushed to the Rechts der Isar hospital to join the other injured, some relatively free from harm but others, like Rayment, seriously ill.

After the last of those rescued had been dispatched to hospital the work of sifting the wreckage for further survivors continued. Some two hours later, after all hopes of finding others alive had gone, a newsman kicked about the debris of the charred remains of the aircraft's tail section searching for a can of film of the Manchester United v Red Star match. Suddenly a piece of cloth was seen to move amongst fragments of luggage, and on clearing the litter young Ken Morgans was found, unconscious but alive and breathing. That brought the total number of survivors to 24, but sadly not for long. News came through that Frank Swift, the *News of the World* reporter severely burned in the accident had died of his injuries. The casualty figure now stood at 21 dead. Of the 23 survivors 15 had been detained in hospital. Six of these were dangerously ill, including Ken Rayment, Matt Busby, John Berry (outside right), Duncan Edwards (left half) and two of the Yugoslav passengers, Mrs Eleanor Miklos and Mrs Vera Lukic.

As airport officials waited at the scene of the crash for the arrival of the German accident investigation team, news of the tragedy flashed around the globe. The city of Manchester was shaken to its foundations. Messages of sympathy were sent from the Queen to the Civil Aviation Minister and the Lord Mayor of Manchester. President Tito of Yugoslavia sent a message to Prime Minister Harold Macmillan: 'I am deeply moved by the news of the disaster which is a heavy blow to British sport and the English people. Allow me to express my deepest sympathy.' The National Union of Journalists stated 'The loss of these distinguished sports writers is a grievous blow to journalism which will be felt keenly throughout the newspaper industry in general and in all walks of life in the north of England.' Indeed, the effects of the catastrophe rippled far beyond northern England and the world of football.

In the hours after the accident the weather worsened and continuous heavy snow fell throughout the evening. The Luftfahrt Bundesamt accident investigation team, headed by Chief Inspector of Accidents Captain Hans-J Reichel, was dispatched from Brunswick in northern West Germany and arrived to examine the wreckage for clues to the cause of the crash. By the time the investigating officials reached the scene it was 22.00hrs local time, six hours after the event, and a thick layer of snow blanketed the twisted metal. Accident investigation at the time was hardly the precise science it is now, but even by the standards of the day the examination seemed somewhat perfunctory. Little or no attempt was made to protect the evidence, and newsmen, photographers, airport workers and officials milled all over the scene. No lighting was available for the investigating team and lamps had to be commandeered from a BBC news cameraman to provide illumination. Reichel's own words from the offical report describe what was found.

'The wrecked aircraft was covered with a layer of snow about 8cm thick. The right wing, which was only slightly damaged, was still firmly attached to the fuselage and had not been exposed to the effects of fire, presented a completely even layer of snow. This was powdery and could be brushed aside with the hand without difficulty. Under this there was a layer of ice, the upper surfaces of which were very rough, frozen firmly onto the skin of the wing. When one ran one's hand over it, it felt like a coarse kitchen grater. The very loose powdery snow lying on top had not blended at all with the layer of ice. It could, for example, be blown off without difficulty, so as to leave the bare layer of ice exposed. I found the same condition at all points on the wing, which I examined thoroughly, with the exception of the part situated above the engine nacelle and in the region of the slipstream. Here, after the snow had been removed, the bare outer skin was visible, without any ice accretion. Apart from this icing I could find nothing which might have been a cause of the accident, or could be considered to have contributed to it.'

And that was that. Icing was the cause of the accident! As far as can be ascertained no further inspection took place, and had it not been for the opposition of others concerned the West German authorities would have disposed of the wreckage the next day.

In London the same evening a BEA accident investigation team was convened and immediately flew to Munich. The Ministry of Transport and Civil Aviation's appointed investigator, G. Kelly, left the next morning (Friday 7 February), but his flight diverted to Frankfurt because of bad weather and his arrival at the scene was later than hoped. At the time of Kelly's departure Captain Thain and the crew reluctantly appeared before a press conference on the promise that afterwards they would be left in peace, a condition to which the press duly consented. Thain, in no fit state to face the rigours of further interviews, was then ordered by the BEA doctor to rest for the remainder of the day.

The West German investigator Reichel, in the absence of Thain, conducted his enquiries at the airport and approached Black, the BEA station engineer, regarding incidents leading to the fatal crash. Most of the discussion centred on the engine problems encountered during the take-off attempts, and then turned finally to details of the conclusive sequence of events.

Reichel: 'What can you tell me of the third take-off run?'

Black: 'The aircraft's nose lifted after it had covered approximately the first third of the runway and then continued in this normal attitude until approximately half or two-thirds of the runway had been covered. I was unable to see whether or not the wheels were on the ground all the time as the aircraft was enveloped in slush and spray during its whole run.'

Reichel: 'Have you any personal opinion on why the accident occurred?'

Black: 'One possibility I feel could be the amount of drag caused by excessive slush on the runway.'

Reichel: 'The captain, of course, should know his aircraft and under what conditions he can attempt to take-off!'

The next morning, Saturday 8th, Captain Reichel obtained an interview with Captain Thain in the presence of the Ministry representative Mr Kelly, BEA's chief investigator, Wg Cdr J. Gibbs, and a number of other officials. After some preliminary inquiries about events leading up to the third take-off attempt, the questioning proceeded to the loss of speed experienced.

Reichel: 'How does Captain Thain explain the drop in speed, if the instruments were reading correctly?'

Thain: 'My opinion is that the aircraft's speed was retarded on the ground, and I think there must have been snow of sufficient depth to retard the speed and not the engines.'

Reichel: 'There were at most four centimetres of snow on the runway, quite wet snow. The impression of the wheels went right through to the concrete.'

Thain: 'When we landed from Belgrade there was a tendency to slide, but very shortly the braking action was positive.'

Reichel: 'The runway is long. The previous take-offs were abandoned in the middle of it.'

Thain: 'My duty, which I performed in the aircraft, is not to look out from inside.'

Reichel: 'When it was apparent that the end of the runway had almost been reached, and the speed was dropping, why did he (Rayment) not abandon the take-off?'

Thain: 'When he reached a speed of 117kt, the length of runway remaining was such that he had no alternative but to keep going. When the speed fell to 105kt and he wanted to abandon the take-off, it was impossible to halt the aircraft before it reached the house.'

Reichel: 'After the accident we established that the starboard wing had a layer of firm and very rough ice on it. Under the covering of powdery snow the surface of ice was quite rough. The wing surface above the engine, in the region of the propeller slipstream, was free from ice. This was established about 30min after the start.' [In fact the inspection took place six hours after the crash.]

Thain: 'When I walked out to the aircraft on the first attempt, I could see the snow thawing on the wings and count the ribs of the aircraft.'

Reichel: 'What do you think was the cause of the accident?'

Thain: 'My personal feeling is that there must have been a large quantity of snow built up at the end of the runway that prevented the aircraft from accelerating.'

Back in England on that Saturday afternoon the decision had already been taken to proceed with the day's sporting events, although of course the United fixture with Wolves had to be cancelled. A two-minute silence was observed at all games, including Football Association and Rugby Football matches. Flags were flown at half mast at all grounds with players wearing black arm bands and the press black ties. Emotions were running high and many team members felt they should not have been asked to play. That evening Captain Thain and his crew flew back to London.

In Munich the weekend was blessed with warm weather unprecedented for the time of the year with clear skies and bright sunshine. A BEA engineering team, acting on behalf of the British Government, carried out a thorough inspection of the engines and reported that, as suspected, no malfunction had occurred. On the Sunday, without further ado, the German authorities, unbelievable as it may seem today, simply sold the wreckage for scrap! The weekend closed with the good news that Matt Busby, although still seriously ill, had been taken off the danger list.

By the morning of 10 February 1958, preparations were complete for the return of those lost in the accident, and before dawn the coffins draped with the Union flag were taken to the airport in readiness for the flight home. That morning the West German authorities, prompted by a number of irresponsible comments which had been appearing in the press over the weekend, issued their own statement which was printed in *The Times*:

'After preliminary investigations the West German Traffic and Transport Ministry finds that the fact that the aircraft did not leave the ground was probably the result of ice on the wings and the captain has not given a satisfactory explanation of why he did not discontinue the final attempt to take off.'

In spite of the fact that the inquiry was still some months off the Germans had been prepared to show an early hand and to indicate quite clearly the directions in which their suspicions lay. The statement was made against all the evidence. Of course, there was still the ice accretion found on the wing by Reichel to be explained, but his inspection had occurred six hours after the accident! The press release came as a tremendous blow to Captain Thain.

On Monday afternoon the dead were flown to England. The aircraft stopped briefly in London for four coffins to be removed before continuing with the remainder to Manchester where they lay at the airport overnight. The next morning the flag-draped coffins were taken in solemn procession through the streets of Manchester to Old Trafford, with an estimated 100,000 crowd standing along the route and outside the ground to pay their respects.

Meanwhile life in Manchester and Munich continued. United obtained a postponement of their match against Sheffield Wednesday from Saturday 15 February to 19 February, and wasted no time in attempting to rebuild the shattered team. In Munich inquiries were continued by the German authorities and the British investigating teams, although along increasingly divergent lines. The Germans were still insisting on the icing theory, in spite of the British observation that slush and melting snow on a cold runway were hardly in keeping with ice forming at the same time on wings which before the incident had been heated for anti-icing purposes on the descent into Munich. Ice on the wings, of course, was the Captain's responsibility and contamination of the runway the Airport Director's, so the conflict of interest was hardly surprising. The problem confronting both teams, however, was that the effects of icing on wings and its inherent danger was a well known phenomenon, while virtually nothing was known of the effects of slush drag. The British Airline Pilots' Association (BALPA) had records of a number of alarming incidents involving slush-covered runways, and was not entirely surprised that a serious accident had finally occurred while attempting take-off from a contaminated surface. A request was sent out for additional reports of similar events and a number of frightening tales came to light. The captain of an empty Viscount on a positioning flight to London sent his account of a take-off from a Manchester runway covered in thick slush. An inspection of the runway was completed by the captain before departure, but in the absence of any guidelines relating to such conditions he decided to take-off.

Above:
Two photographs of the accident scene shortly after the crash. Buildings in the background are still ablaze. The rear section of the fuselage has completely severed aft of the wings.
Via Mrs Ruby Thain

Below:
A clear view of the nose and wing section of this fuselage showing the break of the tail.
Via Mrs Ruby Thain

Below:
The Seffers — father and son — work to free Captain Rayment. Karl Seffer's rubber boots can be clearly seen.   *Via Mrs Ruby Thain*

Bottom:
Snow falls on the wreckage.   *Via Mrs Ruby Thain*

'As I charged down the runway I noticed fluctuations in the acceleration, but was unable to exceed 90kt (V1 would have been about 98kt and V2, 108kt). About two-thirds of the way down the speed decelerated to 85kt, and I hauled the nosewheel off the ground. The speed immediately built up and we unstuck. What was particularly noticeable was the marked surge of acceleration, indicating the large amount of drag which had been on the main wheels, even with the nosewheel clear of the deck. I gathered together (in fact we all did) our shattered nerves, and flew back to London. I certainly appreciated this behaviour of tricycle undercarriage in slush, but due to my lack of experience on the type, did not realise the dangerous drag effect on the nosewheel in particular.'

A number of communications, including one notable report released by the Canadian authorities, had circulated around some airline companies cautioning of the possible dangers inherent in slush, but few had been alert to the seriousness of the charge. One company in particular, KLM (Royal Dutch Airlines), had been sufficiently alarmed by reports to issue brief guidelines to its pilots.

'Take-offs in snow or slush:
  'Do not attempt to take-off in tricycle aircraft, when slush or wet snow is more than 2in deep on the runway.
  'The increased resistance due to slush, which varies according to the speed of the aircraft, may reach such a value that it equals the available thrust and prevents further acceleration.
  'Pitching moment on the nosewheel may also build up so rapidly that it cannot be controlled by the elevators. If the nosewheel is difficult to lift, it is a sign of danger, and the take-off should be discontinued immediately. The elevator trim tab should not be used excessively in such circumstances to help the stick force, otherwise a critical control condition may result as the tail-heavy aircraft suddenly leaves the ground.'

BEA, admittedly in keeping with the knowledge of the time, did not treat slush with the caution we now know it deserves. Operations manuals contained virtually no instructions for pilots on take-off from contaminated runways and judgement was left very much to individual captains; a fact which BALPA, now firmly behind Thain, was at pains to point out to the company.

On Wednesday 19 February a makeshift Manchester United team, including Ernie Taylor bought from Blackpool and Stan Crowther from Aston Villa, with five reserves of whom two were only 17, faced Sheffield before a 60,000 crowd at Old Trafford. United ended the evening by achieving an amazing 3-0 victory and the atmosphere in the city was electric.

Early the next morning, at just after two, Duncan Edwards, probably the greatest football player since the war, died of his injuries in Munich. He was just 23. The casualty figure now stood at 22 dead. In the same hospital Captain Rayment, still in a deep coma into which he had fallen on the evening of the crash, was fighting for his life. At home Captain Thain was also fighting, in this case, for his career. Investigations were leaning much too heavily on the ice on the wings theory as far as the crash was concerned and attempts were being made by

Thain and BALPA to gather more data on the effects of slush. It was even suggested that BEA mount a slush trial using one of the redundant Elizabethans which had already been retired from service to make way for Viscounts, but without success.

As the weeks passed in preparation for the inquiry and the beginning of Captain Thain's ordeal, Captain Rayment's struggle came to an end. He died in Rechts der Isar Hospital on 15 March, the last of those to lose their lives as a result of the accident.

By the middle of April good progress had been made in the preparation of evidence for the forthcoming inquiry. Much detail had been collected in Britain and West Germany and there was growing confidence amongst all participants that the session would be conducted in a fair and just manner. In the Munich hospital all but Matt Busby had been released, but finally, on 18 April, he was declared fit to travel and allowed home. The marvellous work of the staff at the Rechts der Isar Hospital in caring for the injured was gratefully recognised by the award of the CBE to the chief surgeon, Professor Georg Maurer.

The first West German inquiry into an accident involving a foreign aircraft opened at 10.00hrs on the morning of Tuesday 28 April 1958, under the direction of Judge Walter Stimpel, a former Luftwaffe pilot. Reichel, the chief investigator, presented the case on behalf of the West German Commission of Inquiry, whose members were experts in various aviation fields. BEA was also well represented by a number of leading personnel, including their chief accident investigator Wg Cdr Gibbs. Regulations were relaxed to allow Thain to participate, and he and the deceased Rayment were represented respectively by Captains Gilman and Key of BALPA. The sessions were held in private, under guard, in the conference room of a rather stark building at Munich Airport. Despite the gravity of the situation the atmosphere was reasonably convivial.

Reichel began by reading his own report which covered in detail events leading up to and including the final take-off attempt, then turned to the evidence of wheel marks noted on the ground.

Reichel: 'The wheel-tracks from this third attempt were clearly visible at the end of the 1,907m runway and beyond it in the slush. This showed that the aircraft did not become airborne at any time. This was confirmed by most of the witnesses' statements. Just short of the end of the runway the emergency tailwheel is said to have left a clearly visible track for quite a long distance. I myself did not see this track, since numerous aircraft used the runway during the interval before my arrival. The track was however clearly preserved at the end of the runway and on the 250m stopway beyond it, as far as the fence which encloses the airport grounds . . . The tracks of the left-hand wheels were visible only sporadically as far as the fence. The tracks of the twin wheels on the right-hand side of the undercarriage had made a continuous and firm impression. There was no sign of a nosewheel track.'

The emergency tail wheel was so placed to prevent scraping of the tail at high nose-up attitudes and its mark through the slush plainly revealed the extreme nose-up angle being demanded at that stage in the vain attempt to become airborne. Reichel's statement was intended to imply, however, that the excessive nose-up attitude detected at the end of the runway suggested Rayment had been pulling back hard on the control column for some time in an endeavour to take-off, yet something had prevented him doing so. Could the 'something' be ice on the wings? Thain himself, however, had testified that the nose-up angle had not been excessive during the take-off run so the suggestion did not bear close examination.

Reichel continued with the report leading to the discovery of ice on the wings, pointing out that, with the exception of Zulu Uniform, all aircraft which had been exposed to the elements for a long time during their stop at Munich that afternoon had been de-iced before departure. This was a bad point against Thain and was not going to be overlooked by Reichel. Although in the marginal conditions of that day the decision had been taken, correctly as it turned out, not to require de-icing, the fact that everyone else did certainly eroded Thain's standing. The report then turned to the condition of the runway, stating that 'There were varying opinions on the snow conditions of the runway. The quantity of snow which had fallen up to the time of the third attempted take-off would be about 4cm. However, the snow had fallen on a non-frozen and very wet base so that it had subsided to form a layer of slush'. The end of the report contained an account of Reichel's own arrival in Munich, in which it was stated that no problems had been encountered landing on the same runway and in the same direction.

The object of the inquiry was to establish cause, not to put Thain on trial, yet here he was, right at the start, being pushed very much on the defensive. Thain was called to read his statement which was heard in silence apart from one or two minor interruptions from Judge Stimpel to clarify a point.

A number of witnesses were now called including Bill Black, the station engineer, Count Rudolph zu Castell, the airport director, and the young airport trainees who had observed the departure from the top of the terminal building. Surprisingly the experienced controllers in the tower were not summoned to testify, and would not be called independently because Reichel alone selected the witnesses. Written evidence was also presented from a Herr Reinhardt Meyer, the first on the scene of the accident and the witness to the tyre marks through the slush on the runway. Meyer, it was stated, had been unable to attend and was unavailable for cross-examination.

Under questioning at the time of the crash the airport trainees had stated that some snow, especially near the wingtips, was visible before Zulu Uniform's departure and they now confirmed their observations. The photograph taken at the same time showed areas of the wing obscured, and was displayed to support the evidence. Questioning then centred on the attitude of the aircraft, the slush spray generated, and the wheel marks on the ground. The airport trainees each seemed to think the aircraft attitude during the take-off run was higher than normal yet all contradicted themselves by reporting that they saw slush spray thrown up in a bow wave fashion from the front. A number of other witnesses had also observed the large slush bow spray which seemed to indicate the nosewheel actually touching the surface during the run rather than being in the air. This was more in keeping with Rayment's struggle to hold the nose wheel just off the ground against the slush force on the main gear, and the trimming required to ease the strain. Perhaps the nose had dropped inadvertently further down the runway, it was suggested to the trainees? That was difficult to tell was the reply, owing to the amount of spray being thrown up and the distance involved.

Count zu Castell now turned to the inspection of the wreckage and the discovery of ice some six hours after the accident.

Castell: 'During our examination — on many parts of the surface — we quite definitely felt and saw a rough coarse-grained layer of ice on the surface of the wing. The nose of the wing was free of ice, the engine nacelle and the area of the wing behind it were also free of ice though covered with snow . . . I should like to add that there was ice on some parts of the airscrews . . . The rough layer of ice on the wing would have been 5mm at the least, if not thicker.'

No measurement of the ice thickness had been taken at any point so how the figure of 5mm had been arrived at was a mystery. So too was the ice on the propeller blade, which would hardly have been evident during take-off, and must have formed after the accident. However, if ice had formed there since the crash surely the same could have happened with the wings. There was also the mystery of the engine nacelles being free of ice. In the words of Herr Werner Goetz, the airport technical manager: 'I was interested above all in the area near the engines because, in our experience, ice tends to form there first; the engine is warm, the snow melts and then ice forms. But I found no ice under the snow, which was falling continuously'.

The inquiry opened the next day with the reading of various statements taken from other airport trainees who each testified that snow, to a greater or lesser degree, had been seen lying on the wings of Zulu Uniform before departure. This led to a discussion of the weather conditions at the time and a Dr Müller from a nearby meteorological office was called to testify. Dr Müller was careful not to commit himself on any topic, but conceded 'As a meteorologist, one cannot completely exclude the possibility that in the six hours between the accident and the observation of the layer of ice a certain irregular layer of ice could have formed on certain parts of the plane'. Also discussed was the possibility of the half melted snow on the wings freezing as a result of the wind chilling effect during the take-off runs, but this was deemed unlikely. Even if ice could be formed under such circumstances the lack of ice on the engine nacelles, the most probable place for icing to occur, could not be explained.

A Dr H. Schlichting, professor of stream mechanics at Brunswick University, was then called to consider the aerodynamic implications had the ice discovered after the accident been present during the final take-off attempt. He felt sufficiently confident to state that the roughness of the wing surface caused by that amount of ice could make the take-off impossible but was less able to explain in aerodynamic terms the drop in speed experienced. The Munich Airport technical manager, Herr Kurt Bartz, then gave evidence regarding the condition of the runway and details of the drive to inspect the surface carried out on the afternoon of the crash. The time estimated by Bartz to carry out the inspection and the distance travelled (ie the complete length of the runway) did not tally and pointed to a rather inadequate check of the slush, or the use of a very fast vehicle.

He was followed by a Mr J. Kenward, a BEA engineer and a performance expert, who had been involved with the Elizabethan since its introduction into service in 1952. His calculations demonstrated that up to 3in of ice on the wings of an Elizabethan would be necessary to create any noticeable disturbance in the aerodynamic properties. Graphs were also displayed to indicate acceleration characteristics on both dry and contaminated runways showing that wing ice accretion, even to that extent, could not have resulted in any slowing of the aircraft. Slush drag co-efficients were introduced in an attempt to explain the deceleration phenomenon, but since, as Kenward pointed out, nobody knew much about it, the slush drag effect might be much greater than anticipated. The confusion in the inquiry room was now certainly more than expected and amid much perplexity the second day ended.

The final day began with discussions about the roles of the pilots on the day of the accident, and led through a number of points to consideration of decision making on the flight deck. Could any inadvertent misunderstanding have occurred between the two captains, one willing the aircraft to depart, the other to abort? And why was Thain, the commander, sitting in the right-hand seat? No laws had been broken, of course, although BEA did make a point of indicating that company regulations had been broken. And so, in ever decreasing circles, the third day drew to a close. Judge Stimpel summed up by stating that at that moment the commission was in no position to reach a definite conclusion, the scientists were requested to return to their drawing boards, and the inquiry was adjourned until further notice.

On 25 June the inquiry resumed, this time in Frankfurt, in the conference room of the Federal Bureau of Air Safety. Time and further research had not brought the opinions of the scientific antagonists any closer and the original stances were maintained. The same arguments were reiterated except for two notable exceptions outlined by the British team in support of Captain Thain. Mr R. Jones, the British meteorologist, pointed out that 4-5cm of snow would have to fall and melt to produce 5mm of ice. Since it was established that only 0.5cm fell in the 1¾ hours Zulu Uniform spent on the ground, from where had the remainder materialised? Kenward, the BEA performance expert, also propounded an interesting theory. If slush drag were related to speed, not weight as Schlichting contended, then a very different picture emerged. He drew a rough graph of drag against speed showing the effect of retardation of an aircraft if it was considered that increase in drag was related to the square of the speed. The result demonstrated, if his figures were correct, that not only could acceleration be markedly reduced in such cases but that deceleration could actually take place: the exact circumstances surrounding Zulu Uniform's fatal take-off run.

The inquiry closed with a statement from Judge Stimpel:

'(1) It will not be possible to clear up all the events which contributed to the accident. (2) A rough layer of ice on the wing surface undoubtedly impaired the aerodynamic properties of the aircraft. This layer contributed substantially to the accident. (3) It is not impossible that other circumstances may have contributed, which it is no longer possible to determine in detail.'

The official report was to follow.

In the meantime Thain turned his attention to the poultry farm and struggled to maintain his sanity while in the football scene Manchester United strived to regain their poise. They were, perhaps not surprisingly under the circumstances, knocked out of the European Cup by Milan, but by a magnificent effort managed to reach the final of the FA Cup where they were defeated by Bolton Wanderers. So ended the 1957/58 football season. The summer months went by, the next soccer season began, Christmas passed, and the New Year dawned with still no statement from the Germans. Finally, on Friday 9 March 1959, the long awaited report was released. The layer of ice on the wings had been found to be the 'decisive cause' although not the 'sole cause' of the accident. The inquiry, and Thain's career, had come to an end.

Following publication of the report the BEA Air Safety Committee was quick to follow with its own conclusion.

'Cause: (1) The accident was due to the aircraft failing to reach the required speed to enable it to become airborne in the conditions prevailing. (2) The German Commission of Inquiry had attributed this to icing of the wings, which had been described in their Report as "the decisive cause". (3) The Committee feels unable to accept this evaluation of the importance of icing, but accepts that it was certainly a significant factor. Slush on the runway may have been another important factor, and there may also have been other contributory causes. (4) The Committee feels it is not possible to evaluate the exact degree of importance attributable to these two factors, either singly or in combination. (5) The Committee notes that at the time of the accident Captain Thain, designated Captain of the aircraft, was not occupying the left-hand seat, thereby contravening Flying Staff Instructions. (6) The Committee notes that the aircraft was not de-iced at Munich.'

In due course Thain's flying licence was suspended by the Ministry in a manner considered somewhat controversial, and questions were asked in the House of Commons regarding his treatment. Meanwhile Thain, unwilling to accept the report's findings, sought further evidence to reopen the inquiry and in the process turned over some very interesting stones. Black, the BEA station engineer at Munich, traced some local engineers who worked for Pan Am. One Otto Steffer, and his son Karl-Heinz, who also worked at the airport as a part-time fireman, had helped free Captain Rayment from the wreckage. Black obtained a statement from the younger Steffer declaring that he had walked over the wings inboard of the engines to climb on top of the shattered cockpit and had found no evidence of ice deposit in a place where Reichel had discovered ice six hours after the crash.

Analysis of the Munich runway camber showed that slush ridges could have trapped water to a depth of as much as 4-6in at the runway edge, and corroborated pilots' accounts of large puddles forming at times on the runway. Mrs Thain, a graduate chemist, came up with the answer to the mystery of no ice on the wings behind the engine nacelles: fire extinguisher discharge directed at fires mostly in the vicinity of the nacelles would have covered the wing surface in the area with an effective anti-freeze which would have prevented icing. It was also discovered that due access to all documents had not been achieved and the West German authorities were now approached with a request to supply all material. Inspection of the data revealed, to an amazed Thain, that not only had the air traffic controllers not been invited to attend the inquiry but vital evidence submitted by them had not been presented to the commission. In the words of one controller:

'It (Zulu Uniform) began rolling normally and built up speed until it was about halfway along the runway: the nose wheel left the ground, but touched down again after about 60-100m.'

Here was direct evidence of problems in maintaining the nose-up attitude associated with slush, yet, unbelievably the information appeared to have been suppressed.

All the new evidence was presented with a request to reopen the inquiry, but to no avail. On 14 March 1960, the West German authorities announced that 'the facts, evidence and other points to which the attention of the commission was drawn do not justify the reopening of the proceedings'.

The authority of the Ministry of Aviation to revoke Thain's flying licence without due investigation on its own part was called to question and on Monday 4 April 1960, a hearing was opened to review whether Thain's duties as captain of Zulu Uniform had been properly performed. The hearing was not assembled to assess the cause of the accident, although, of course, the event would feature prominently in discussions. After four days of deliberation the review concluded that Captain Thain had not conducted his duties properly in assessing whether the wings were clear of ice and snow. It was another tremendous blow to Thain, but more was yet to come. The report was not made public until 12 October, over 2½ years since the accident, during which time Thain had been grounded by BEA pending the outcome of the various sessions. The Ministry conceded that his time away from flying exceeded any period of suspension that might have been imposed and duly offered the return of his licence. The following day Thain received a letter from BEA stating that since company regulations had been broken by his failure to occupy the left seat and to assess properly the condition of the wings of his aircraft, his employment was now terminated. At this point Thain's morale could sink no lower; from here the only way to go was up.

Fresh slush trials were conducted in the United States with new evidence being amassed, but further approaches to the Germans to open the inquiry were simply rebuffed. Manchester United's court case against BEA alleging negligence was imminent, but suddenly, without warning, it was dropped and settled out of court for about one tenth of the original sum claimed. At least one of the aerodynamic experts preparing evidence for the case stated, 'The effect of the deposit on the wings was at the most of marginal significance and then only when conditions were already very critical owing to slush'.

The course of events was beginning to turn.

Scientific interest in the effects of slush was also on the increase in Britain, and by the end of 1963 trials of a number of aircraft, including the Elizabethan, were begun at the Royal Aircraft Establishment, Farnborough. The results were not published until April 1964, but clearly showed that all suspicions regarding the effects of slush had been confirmed. The theory proposed by Kenward, the BEA performance expert, that slush drag was related to the square of the speed was proved accurate and the tests demonstrated quite clearly that with a depth of as little as 0.5cm of slush the Elizabethan required 50% more runway for take-off than when dry. Now with the backing of the Ministry of Aviation the Germans were asked to consider the latest evidence and invited to reopen the inquiry. A further year and a half was to pass before the sessions opened on 18 November 1965, to re-examine the case. Thain now faced the inquiry convinced that justice was, at last, about to be done, but his hopes were not to survive for long. The West German authorities were adamant that no new witnesses would be called, although, of course, they were prepared to listen to the additional evidence regarding slush. After two days of discussion the re-opened inquiry drew to a close with the result a foregone conclusion. A draft of the report sent to Britain clearly indicated that, 'Icing was still to be regarded as an essential cause of the accident . . . slush, however, must be regarded as a further cause . . . the command structure on board G-ALZU was not entirely clear-cut and may also have had an unfavourable effect'.

There were so many errors and inconsistencies in the documents that the British baulked at the distorted facts and implored the Germans to revise their findings, but they would not be moved. By April 1966 the Ministry of Aviation, tired of the German recalcitrance, issued, to its credit, its own deduction.

'We conclude that there is a strong likelihood that there was no significant icing during the take-off . . . we conclude that the principal cause of the accident was the effect of slush on the runway.'

In spite of the Ministry view the West Germans were not to be swayed. However, thanks to a timely comment from the then Prime Minister, Harold Wilson, by coincidence on a visit to Manchester United at Old Trafford, a new British hearing was finally ordered. The sessions opened in London on Monday 10 July 1968, more than 10 years after the crash. All previous evidence was to be reviewed and all available witnesses interviewed or their written statements considered. It was to be a most full and thorough investigation.

The key witness was to be Herr Meyer, the first on the scene of the crash and the observer of the marks on the runway. He had been unable to appear at the original hearing, or so it was assumed, and had not yet testified personally. Meyer was then living in the United States but at the time was on holiday with his family in Bremen, and members of the commission travelled there for the interview. Not only was Meyer an aeronautical engineer but also a trained pilot, and his comments could be received with the widest respect. What he had to say, however, came as a surprise to all concerned. He had examined the wings of the stricken aircraft shortly after the crash and had noted that 'there was nothing like frost or frozen deposit: that I know definitely. There was melting snow only'. He continued that he had imparted this information to Reichel but had heard no more of it. And why hadn't he presented this information to the inquiry? 'I was not called to give evidence and I didn't even see them' was Meyer's reply. He had been deliberately omitted, along with a number of other key witnesses, from the German inquiry. It was incredible! It was also revealed that there had been tampering of Meyer's original written statement and that his comment on the absence of ice on the wings had been removed. A summary of the irregularities in the presentation of evidence in the German Commission of Inquiry declared that 'Meyer's statement was truncated and the most significant part of his evidence was concealed'.

The mystery of the photograph purporting to show areas of the wing surface obscured by particles of snow was also solved. The original print had been made from a copy negative, but now a fresh print from the original negative had been acquired and clearly showed the lettering and marking on the surface of the wings. Photographic experts were called to examine this new evidence and unanimously agreed that in their opinion there was no sign of ice or snow on the wings.

The report of the proceedings was made public on 10 June 1969, and made interesting reading.

'We are satisfied that at, or after, reaching V1 the aircraft's nose-wheel re-entered the slush . . . had (this) not happened, the aircraft must have flown off . . . We are equally satisfied that the descent of the nosewheel was caused by increased drag exerted through the main wheels, in other words, by the aircraft entering a trail of deeper and/or denser slush . . . This increase in slush drag . . . is in our view the prime cause of the accident. We are satisfied that thereafter the aircraft ran with six wheels in the slush until towards the end of the runway it rotated to the point where the tailwheel made contact with the ground . . . We can be sure that the period between V1 and rotation was one of deceleration . . . We cannot reach certainty as to the speed at the end of the period . . . Once the aircraft was rotated without lifting off, the accident was inevitable. There was not enough runway left to regain sufficient speed . . . Our considered view, therefore, is that the cause of the accident was slush on the runway. Whether wing icing was also a cause, we cannot say. It is possible, but unlikely. In accordance with our terms of reference we therefore report that in our opinion blame for the accident is not to be imputed to Captain Thain.'

To this day the West Germans still refuse to acknowledge the findings of the British commission.

As the dust settled, Captain Thain carried on running his poultry farm, as well as farming some 30-odd acres of land in Berkshire. He disbanded the poultry business in 1969 and continued farming until the strains of his earlier struggles took their toll and he died of a heart attack at only 54 years of age in 1975.

747 Ice/Snow Max Depth Data

Dry Snow		*Wet Snow/Slush*
1½in (38mm)	— Take-off —	½in (13mm)
4in (10cm)	— Landing —	½in (13mm)

Below:
Original print from copy negative apparently showing snow covering the lettering on the top surface of the wings. *Via Mrs Ruby Thain*

The Trident Tragedy 1972

To strike or not to strike, that was the question: or, more correctly, the argument. At London's Heathrow Airport, the atmosphere in British European Airways (BEA) crew room, like the weather outside, was changing for the worse. Amid the normal bustle of activity in Queen's Building on a busy Sunday afternoon in June a large number of pilots had formed an impromptu gathering to discuss their present dispute with the company. Some flight crew present were going to or from work while others, like Captain Stanley Key, were on standby duty in case of disruption to rostered crews.

BEA and the British Airline Pilots' Association (BALPA) had for some time been at odds over a large number of issues which centred on pay and conditions. The dispute had not only created a certain amount of acrimony between the company and its flight crews but had also divided the pilot community. Many pilots were vehemently opposed to industrial action while others strongly favoured a call to strike. One pilot group of 22 supervisory first officers had already withdrawn their services causing more than a little embarrassment to BEA. Long periods of supervisory duty and poor monetary rewards were not felt to justify the supervisory pilots' own lack of aircraft handling.

In previous years a large recruiting campaign had been conducted by British airlines, and BEA in particular was still active in processing its young pilots through basic training. After completing the commercial pilot's licence and instrument rating at air training school, recruits joining BEA underwent an intensive course of instruction and training on the aircraft to which they were assigned. Ground school, simulator and base flying programmes were undertaken and all had to be completed satisfactorily for licence requirements. Further training and experience was then gained 'down line', ie on BEA's normal commercial services, under the strict supervision of suitably qualified flight personnel, before the recruits were deemed competent to operate unsupervised.

The Trident aircraft was crewed by three pilots with the captain (P1) in the left-hand seat, the co-pilot (P2) in the right-hand seat, and another co-pilot, in the position of third pilot (P3), seated behind the other two just aft of the centre console. Under normal operating circumstances the P1, and whoever occupied the P2 seat, would share clearly defined aircraft handling roles and co-pilot tasks, at the captain's discretion, while duties of the P3 included operation and monitoring of systems, completion and checking of paper-work, and monitoring of flight progress. P2 and P3 would also alternate duties by changing seats as agreed with the more senior pilot normally taking precedence. Whilst under training as P2 or P3 the recruits were supervised by a training captain in P1 seat, with the addition during P3 training of a supervisory first officer sitting in the jump seat, an extra fold-away seat fitted on all flight decks for such contingencies. During such P3 training duties an experienced co-pilot occupied the P2 position. The system was tried and tested and produced competent, well qualified pilots.

One young man present in the crew room that Sunday afternoon on 18 June 1972, was 22-year-old Jeremy Keighley. A keen and enthusiastic pilot, he was one of 36 trainees unable to complete his 'down line' experience because of action by the supervisory first officers. He had finished his P2 training and was qualified as co-pilot, but had flown only one sector of line training as P3. Having already completed 16hr 20min P3 training on the simulator and 6hr 45min base training on the aircraft, it was most frustrating for this young man and his similarly inconvenienced colleagues to be denied the chance of finishing their training. BEA was none too happy about the situation either, and was faced with rostering difficulties as a result. To resolve the dilemma, the airline decided to continue using the trainees for flight crew duties by simply restricting their activities to those for which they were qualified, that is, as co-pilot sitting in the P2 seat. The anomaly therefore arose that when 'P2 only' pilots were rostered to fly unsupervised they could only occupy the P2 seat, despite being the least experienced of the crew. Thus the junior of the two co-pilots, some with only 250 total flying hours and about a dozen hours simulator and aircraft handling practice on the Trident, would be the pilot required to take control in the event of incapacitation of the captain. To help maintain continuity of experience a 'brown line' system was introduced whereby pilots of less than 12 months' experience on 'type' had their names underlined by a brown line and were required to fly with a qualified co-pilot of more than 12 months' experience. Even so, two junior second officers, one relatively inexperienced and one very inexperienced could be rostered to fly together. Although the young men restricted to P2 duties were acknowledged to be enthusiastic, reliable and well trained, deep concern was expressed by a large number of captains over their lack of experience which was felt to be far too little for safety. Earlier that same month *Aerospace Magazine* had published an article written by airline Captain R. C. Leighton-White who supported these views and which stated 'to critically monitor and question a captain's actions the co-pilot must have sufficient training and experience to have the necessary confidence in his own judgement to question a senior pilot's actions'.

On the previous Thursday (15 June) a certain captain had been rostered with one of the 'P2 only' co-pilots to fly to Dublin and back and then on to Nicosia. Aware of the below-average facilities at Nicosia and the possibility of rapid deterioration of the weather, the captain made known his concern and asked for a fully-qualified replacement. His request was denied. Upset by the rebuff and the attitude of management he vented his feelings on the young co-pilot and more or less told him that he would be useless in an emergency. The outburst had a very upsetting effect on the P2 and did nothing to improve his confidence. On the departure out of London for Dublin the call to select 'flaps up' was made as normal at noise-abatement time, but, still flustered by the incident on the ground, the young man instead selected the flaps fully down. The P3 immediately spotted the action and quickly reversed direction of the flap lever before movement of the flaps. On the flight deck events happened quickly, and it takes training, practice and experience to function normally in what is potentially a hostile environment. For the new recruit, with as yet undeveloped skills, a great deal of concentration is required to avoid mistakes and the occasional lapse is, perhaps, not surprising. The pressures on these young men, albeit of high calibre, can be very great.

The young P2 involved in the 'Dublin incident' happened to share a house with Second Officer Jeremy Keighley who was quickly made aware of the story. The two met amidst the group of pilots gathering in the crew room that Sunday afternoon and Keighley, on standby duty at the same time as Captain Stanley Key, enquired of his friend what Key was like to fly with. The other P2 had no personal experience of working with Stanley Key although the Captain's name was well known amongst Trident crews because of a somewhat unfavourable reputation. One of the more puerile aspects of the dispute was the appearance on Trident flight decks of offensive graffiti and Captain Key, known as one strongly opposed to strike action, had been singled out for a certain amount of abuse. Whatever the criticisms, however, his flying skills were generally acknowledged and he was recognised as a very competent pilot.

A confidential postal ballot had been arranged by BALPA in an attempt to resolve the dispute and discussion amongst the pilots in the Queen's Building crew room focused on the outcome. Strong opinions were expressed by both sides as to their preference. As votes were being received at the Association headquarters adjacent to Heathrow, Captain Key, on his own initiative, had been attempting to canvass support against strike action amongst his senior colleagues. He was naturally guarded as to the results of his efforts and was strongly opposed to discussing his actions amongst the pilot community at large. In the rather heated atmosphere at the time one first officer made the mistake of asking how Key's endeavours were progressing. The result was instant and furious. Captain Key completely lost his temper and a raging outburst ensued. The first officer was told in no uncertain terms that such matters were none of his business, or anyone else's, and that the matter had the right to be afforded the same confidentiality as the BALPA ballot. In the eyes of at least one witness it was 'the most violent argument I have ever heard', even if it was somewhat one-sided. Captain Key's anger, however, abated as quickly as it had flared. To his credit he took the first officer in question by the arm and in front of the same group apologised for his behaviour. Thereafter his manner seemed normal. Second Officer Keighley was also a witness to the spectacle and the effect must have been more than alarming. Like all new recruits he was aware that BEA had its own small number of difficult flight crew members, just like any other airline. Whether Captain Key fell into this category or not is certainly open to conjecture, although he was renowned for a somewhat brusque manner. He was a man of traditional values, of Royal Air Force and airline service. The new entrants to the company from the air training schools were of a different breed. They had no military training or airline flying experience and were quite alien to Stanley Key. They were of another generation, some 30 years younger. He had little in common with these new entrants and was somewhat suspicious of their abilities. A few, like Jeremy Keighley, were very raw recruits indeed. It would not be unfair to say that Captain Key probably felt as uncomfortable in the company of these young men as they did in his. After the incident in the crew room Keighley's impression of Key is not known, although the effect could hardly have been reassuring. The effect of the argument on Captain Key, however, was far more sinister.

Captain Stanley Key was short in stature and, at 51, seemed to have no more than the usual problem with weight for a man of his age and build. To those who knew him he presented a 'picture of robust good health'. He actively pursued his gardening hobby and neither his wife nor family doctor had any reason to suspect anything amiss. However, something was very seriously wrong with Captain Key's health and his condition had been progressively deteriorating over a very long period. As far back as his 20s he had been suffering from atherosclerosis. The disease had now reached a severe stage and the arteries of his heart had been narrowed in places by some 50-70%. Unknown even to those close to him, his life expectancy was short.

All flight crew, of course, undergo medical examinations at regular intervals, which for a pilot of Key's age would be every six months with the addition each year of an electrocardiogram (ECG). This test reveals any heart muscle abnormality but, unfortunately, it rarely shows narrowing of the arteries. As a result, neither of Key's previous two ECGs, in 1970 and 1971, indicated a problem. It has been argued that a 'stress ECG', where the patient is subjected to rigorous exercise before readings are taken, may have detected the diseased arteries, but such tests were not considered by medical opinion at the time to be sufficiently reliable. Too many 'false positives' and 'false negatives' resulted. Captain Key's heart condition could have been ascertained by means of an arteriogram, but the procedure was deemed much too risky for routine examinations. To all intents and purposes, therefore, Key's regular check-ups proved him to be a fit and healthy specimen.

The fatty or fibrous deposit on the walls of Key's coronary arteries escaped detection and remained a serious health risk. Within the pathologically thickened walls small blood vessels developed, which, in keeping with the unnatural growth, were weak and prone to rupture. During Captain Key's very angry outburst in the crew room his blood pressure soared to dangerous levels with damaging results. At this moment, as near as can be certain, a delicate blood vessel in the left coronary artery ruptured causing haemorrhage. The resulting pressure from the blood on the arterial wall would, with time, force separation of part of the arterial lining. At that juncture the effect would be 'anything from slight pain akin to indigestion at one end to nigh death at the other'. It would certainly not be a moment of concentration and alertness.

About one hour after the altercation in Queen's Building, a standby crew was 'tannoyed' for service. The crew rostered to fly the BE548 to Brussels was delayed on their return to London and others were required to take their place. The crew selected consisted of 51-year-old Captain Stanley Key, unknowingly suffering from a serious heart condition; 22-year-old Second Officer Jeremy Keighley, of little experience and restricted to 'P2 duties only', and 24-year-old Second Officer Simon Ticehurst, competent, conscientious, and relatively experienced, but still only a junior officer.

Captain Key had 15,000 flying hours to his credit, with 4,000 hours in command on Tridents. His skill and ability could not be questioned. On proceeding to the aircraft, however, Key would be experiencing a degree of discomfort leading to increasing pain. The effect would be sufficient to distract his train of thought and progressively to affect his reasoning powers. The ultimate tearing of the arterial lining would not be far away, with serious consequences awaiting. If such an event occurred during a critical stage of the flight, control would rest in the hands of a young man who, through no fault of his own, had very little experience. Keighley would doubtless also be feeling ill at ease on the way to the flight, but what apprehension he felt at the encounter with Key can only be surmised. Reports of his progress through basic and advanced training indicated that, when judged by the very high standards demanded, he was somewhat underconfident. On the simulator he was found to be 'slow to react to an emergency' and that 'he lacked initiative'. With determination and a desire to succeed he had progressed steadily through each stage and there was little doubt that with time 'he would make a good, reliable pilot'. However, a handing-over report the month before had reported him as 'slower than average and will call for patient, rather than pressure handling'. What kind of handling could Second Officer Keighley expect from Captain Key this day?

Monitoring the flight and the other pilots' performance would be young Ticehurst, a thoroughly reliable and able pilot. He already had a total of 1,400 hours under his belt, with 750 hours on Tridents and would carefully chaperon their progress.

As the three standby pilots mounted the steps of their Trident aircraft, the weather continued to worsen, turning the day into a typical wet British Sunday. In the late afternoon an approaching cold front, now lying some 30 miles to the west, resulted in overcast skies with a cloud base of 1,000ft, fresh winds from the southwest, and rain. Flying conditions were unpleasant and turbulent. The Trident I being boarded by the crew, G-ARPI (Papa India) was, as the type name suggests, a three-engined aircraft similar to, although smaller than, the Boeing 727 which it preceded in conception. The first Trident I flight took place on 9 January 1962, and the first Boeing 727 flight more than one year later on 9 February 1963. The Americans very quickly caught up, however, and delivery flights of the Trident I to BEA and the Boeing 727 to United Airlines were completed in December and October 1963 respectively. Both types had the three engines grouped at the rear with the tailplane placed free of the engines high on the tail fin. The resultant configuration had certain advantages over wing-mounted engine designs, the simplest and most obvious being the ability to instal an odd number of engines. Asymmetric thrust (ie one engine failed) problems were also reduced, thus allowing take-offs in low visibility. On wing-mounted engine aircraft, failure of an outboard engine on take-off can result in a pronounced swing because of the large power imbalance. Better visibility is therefore required for pilot guidance in case such an event occurs. The advantages of the tail-engined jets, however, are not achieved without penalties. Severe engine failure on a tail mounted design increases the possibility of damage to another engine because of their close proximity, and, with most fuel contained in the wings, fuel pipes running through the fuselage are a potential fire hazard in the event of an accident. Trident and Boeing 727 type aircraft also have the potential to 'deep stall'.

As an aircraft slows, to maintain lift the angle of the wing to the airflow has to be increased by raising the aircraft's nose. If the speed is allowed to become too slow and the nose-up attitude is excessive, a point is reached at which the smooth airflow over the wing breaks down and most lift is lost. This condition is known as 'stalling'. With lift lost from the wings the aircraft flutters flatly from the sky like a falling leaf. Recovery is achieved by forcing the aircraft into a dive using full down elevator and by applying full power until flying speed is once again achieved. The aircraft can then be pulled out of the dive and flown straight and level. Stalling can also occur with an inadvertent change of the wing configuration resulting in lift loss. On modern jets the wings are swept back at a large angle to allow aircraft to cruise at high speeds by delaying the onset of shock waves as the airflow over the wing approaches the speed of sound. At slow aircraft speeds, however, the lift producing qualities of the wing are poor. High lift producing devices are required to improve lift by increasing the surface area and by altering the camber of wing profile. On the Trident I aircraft a combination of Krüger and droop systems were employed at the wing leading edge, with conventional flaps extending at the trailing edge. The droop system, peculiar to Trident I aircraft, increased the camber of the wing by quite literally drooping the leading edge of the wing by hydraulic operation. The droops were hinged at the bottom and when extended moved outwards and downwards from the top, the resultant gap between the wing upper edge and the rear of the droop being closed by a sealing plate. Inboard of the droops, Krüger flaps were hinged along the leading edge and were extended downwards and forwards like a narrow door along the inboard wing length. The droops extended over the larger part of the wing and as such the leading edge droop and Krüger devices were simply referred to as droops. The Trident was the first British aircraft to be fitted with retractable leading edge high-lift producing devices and the system was simple and effective, but not without its problems. The droops and flaps were operated independently by two separate levers situated side by side to the right of the centre console. The lift producing qualities of the droops were considerably greater than the flaps and the danger of inadvertent retraction of the droops instead of the flaps at low speed was readily apparent. (Aircraft since have a single lever for both leading and trailing edge lifting devices.) If the droops are retracted unintentionally at too low a speed and the high lift generated by these devices is lost, the aircraft will instantly be close to stall condition and in imminent danger of falling from the sky. In a stall caused by this change of configuration, re-extension of the droops would be the obvious recovery procedure, although full stall recovery action may also have to be initiated. To prevent inadvertent retraction of the droops instead of the flaps the droop lever was protected by a mechanical guard throughout most of the flap lever range. After selection of flaps up, the droop lever was unguarded. A climb to 3,000ft, and further acceleration to the minimum 'droops up' safety speed of 225kt, was then required before droop retraction. During this period of about 2min the droop lever was unprotected, but since there was no requirement to operate either lever a speed guard was not considered necessary. At a later date a speed guard was introduced. An amber warning light placed forward of the droop lever was arranged to illuminate if the droop lever was out of position: ie if the airspeed was too low when retracted or if excessive when extended.

96

On a more conventional aircraft at the time, approach to the stall was announced by buffeting of the aircraft as air over the wing became turbulent. If no action was taken the aircraft entered the stall and the nose pitched sharply down, thus aiding recovery. On the Trident there was no buffeting at approach to the stall with droops extended. On T-tailed rear engine aircraft, design characteristics are such that at the stall the nose tends to pitch up, thus exacerbating the situation. If the stall is allowed to develop the engines drop into the turbulent flow from the wings. The disturbed airflow causes engine 'hiccup', known as a surge, resulting in loss of power. If the stall develops even further the high tailplane drops into the turbulent flow and elevator control becomes ineffective. The aircraft is now said to be deep-stalled and, with no means of reversing the situation, recovery is virtually impossible. The inherent danger of such a condition was made apparent as early as 3 June 1966 when Trident I G-ARPY deep-stalled on a trial flight during stall tests. The aircraft fell flatly from the sky and pancaked onto the ground killing all the test crew on board. For this reason Tridents, as well as other British aircraft with tail-mounted engines, have always been fitted with stick pushers as well as stick shakers.

Pilot training covered thoroughly the recognition of, and recovery from, an early approach to the stall condition long before stalling actually occurred. The Trident I was also well protected with automatic stall warning and recovery devices including a 'stick shaker' which physically shook the control column at an early stage approaching the stall condition, and a 'stick pusher' which delivered a hefty push forward to the control column at later stall development. On sensing approach to the stall condition, incidence probes on each side of the fuselage would activate electric motors which vibrated the control columns. At stall onset a pneumatically controlled ram would force the column forward before further stall development. Positioned by the airspeed indicators was an amber 'stall recovery operate' lamp which illuminated during function of the stick push and a red 'stall recovery fail' lamp which lit up with system failure. An amber 'low pressure' warning light, situated forward of the droop lever by the 'droop out of position' warning light, was also arranged to illuminate if the integrity of the stick push system was affected by reduction in air pressure below a certain level.

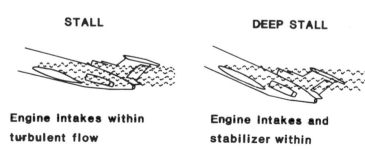

STALL	DEEP STALL
Engine Intakes within turbulent flow	Engine Intakes and stabilizer within turbulent flow

Above:
The problem of stall and deep stall.

Both stall warning devices had encountered teething troubles in the early days and a general if perhaps inaccurate feeling had been left amongst pilots that the systems were unreliable. False warnings had occurred causing genuine concern. It is not difficult to imagine the alarm that might be generated by spurious operation of the stick push during take-off or landing. As a result if the stick push operated in a situation which seemed doubtful to the crew the tendency was to disconnect the stick pusher even though the warning might be genuine. To inhibit the systems, circuit breakers could be pulled out to cut off power to the electric motors and a 'stall dump' lever, situated on the left rear side of the central control pedestal, could be selected to deactivate the stick pusher ram by discharging the compressed air outboard.

A number of incidents over the years had highlighted the dangers of inadvertent flap/droop selection and served to emphasise the concern. Normal procedure was to select flaps up at noise abatement time, which varied according to the circumstances, but was usually about a minute and a half after brake release. The thrust levers would then be retarded to reduce power to the noise abatement setting. Exceptions to the rule included departures from airports such as Rome where take-offs over the sea did not require noise abatement procedures and on these occasions pilots often retracted flaps early to improve climb. A few captains, however, were known to employ the technique more frequently than was permitted, and on a 'small number of occasions' flaps had been recorded as being selected up immediately after landing gear retraction at heights *below* 50ft.

With only the droop lever remaining to be selected up, the normal 'flap up-power reduction' sequence was broken. At noise abatement time, the non-handling pilot could then by force of habit 'select the only lever available and inadvertently retract the droops'. With the speed, at this stage, some 60kt below the minimum droop retraction speed the aircraft would instantly enter the stick push stall régime.

A voluntary and anonymous reporting scheme introduced within BEA revealed an incident which occurred to a Trident departing Paris, Orly for London, Heathrow in December 1968. The first officer handled the aircraft whilst the captain performed the co-pilot functions. Just after raising the landing gear the captain, without saying anything, selected the flaps fully up to improve climb performance. At noise abatement time the inevitable happened. The captain retracted the droops and retarded the thrust levers to noise abatement climb power. The aircraft immediately sank rapidly from the sky. Fortunately the first officer was aware of the captain's reputation and was alert to the situation. He was still hand flying the aircraft and simply pushed the nose down to gain speed as he saw the droop gauge move. Aware of his error the captain re-extended the droops. The prompt action prevented operation of the stall warning devices and the aircraft quickly regained its normal flight profile and proceeded to climb steadily away.

In May 1970 a similar event befell a Trident II aircraft bound from London to Naples. The Trident II was, in fact, provided with leading-edge slats instead of droops but their effect was similar and pilots tended to refer to them as the latter. The incident occurred between 1,200ft and 1,400ft, in good visibility and at an indicated airspeed of 175kt. The captain was handling the aircraft and at noise

98

abatement time the first officer carried out the procedure. After power reduction, with the P2's hand just withdrawing from the thrust levers, a stick shake occurred followed very quickly by a stick push. A scan check of the situation by the crew revealed no apparent problem. A few seconds later another stick shake and push ensued in quick succession. Airspeed and attitude seemed normal and the aircraft was still accelerating. The crew could see the ground and the condition appeared safe. Since no warning lights seemed to be illuminated the captain assumed the stall warning devices to be malfunctioning and called for the stick push to be dumped. On turning to activate the stall dump lever the first officer noticed the droop gauge indicating up and he immediately re-extended the droops. During this period the aircraft 'just about managed to stay flying'.

.At the subsequent investigation each pilot denied selecting the droops up, yet extensive examination of the systems revealed no evidence of mechanical malfunction. One of the crew must have operated the droop lever. Strong but inconclusive evidence suggested that the captain involved in this event was the same person in command when the incident occurred departing Orly. It was difficult to avoid the conclusion that the fault lay once again in the practice of non-standard procedures and that the 'flaps had been retracted early after take-off and then the droops had been retracted at noise-abatement time in mistake for the flaps'. These two incidents, more than any other, served to highlight the dangers inherent in inadvertent retraction. The maintenance of proper procedures was strongly emphasised by management, although any threat of disciplinary action by the company was deemed unwise in the hostile industrial environment.

With time the matter faded into the background. Little or no information was circulated on the effects of change of configuration on stall warning and stick push function; few were aware that in such circumstances, with droops retracted early, stick shake and stick push were almost coincidental.

Captain Stanley Key operated his aircraft by the book. Being a route check captain he was aware, more than most, of the importance of adhering to correct procedures. He liked the use of the autopilot, however, rather than hand flying, and he tended to engage it early, although the practice was not unacceptable. As he settled in the left-hand seat of the Trident in preparation for departure, the discomfort in his chest must have been becoming more intense. Rain from the worsening weather spattered the windshield in front of him and the aircraft rocked occasionally in the gusty wind. The time was now about 16.20hrs BST (15.20hrs GMT) and, with scheduled departure only 25min away, a busy time lay ahead. The aircraft they were checking, Papa India, was about six years old and had been involved in an incident four years earlier while on the ground at Heathrow. An Airspeed Ambassador belonging to BKS Air Transport suffered failure of a flap rod on the left wing on final approach to land. The aircraft banked steeply to the left out of control, turned towards the terminal buildings of the central area and tore the tail off Papa India before crashing into another Trident I parked nearby. All on board the Ambassador were killed. £750,000 was spent on Papa India to return it to flying condition and the marks of that incident could now no longer be seen by the naked eye. On the flight deck however, marks of graffiti on the P3's desk were visible for all to see. As Second Officer Ticehurst

Above and Above right:
Flight deck with detail of the pedestal and throttle controls. *BEA Via John Stroud; BEA*

Below:
Trident leading edge droop. *Via John Stroud*

completed his checks he could not have missed the unpleasant comments directed at his captain sitting only a short distance away.

BEA's scheduled flight BE548 (callsign Bealine 548) from London to Brussels was to depart with each of the 109 seats taken. A world-wide 24-hour strike of pilots had been called by the International Federation of Airline Pilots' Associations (IFALPA) for the next day as a protest against the failure of governments to deal with the problem of hijacking. The effectiveness of the strike could only be surmised, but many passengers were taking no chances and were flying that Sunday evening to avoid possible disruption. A few intending travellers who had not booked tried several times in vain to secure a seat. A maximum payload of passengers and baggage, plus a fuel requirement of 8,200kg resulted in a take-off weight of 50,000kg. The low fuel figure for the short flight left the Trident still some 2,000kg short of its maximum take-off weight of just over 52,000kg and, if necessary, more fuel could be loaded in the wings, but not another kilo of payload could be carried.

As the last of the passengers boarded the aircraft, the crew continued with their checks. Systems and equipment were tested, allowable defects examined, fuel figures scrutinised, and engine power settings and take-off speeds calculated. In spite of the thoroughness of procedures, however, one hidden fault remained. Unknown to the crew a locking wire was missing on the three-way valve of the stick push system. A jolt in turbulent flying conditions might just be enough to move the valve and upset the integrity of the device. With power reduction at noise abatement time any slight misalignment of the valve might effect a small pressure drop. Although the system would not be rendered inoperative, the condition might illuminate the amber 'low pressure' warning light situated forward of the droop lever if the system pressure dropped sufficiently low.

Captain Key was to fly the aircraft and he briefed the crew thoroughly on the departure routeing. The departure clearance had not been received, but from the flight logs and knowing the runway in use, the crew could speculate on a Dover One standard instrument departure (SID) and could plan accordingly. The Dover One SID required the aircraft to pass over a marker beacon situated about a mile or so off the far end of the runway, then to turn left over the town of Staines and track towards the radio beacon at Epsom. The Epsom beacon was to be crossed at 3,000ft or above and the first height restriction was 5,000ft. The drills required in the event of an emergency on take-off, and any other relevant details were discussed. Each man was to carry out his duty as per the flying manual: the captain would operate the radio as well as fly the aircraft, P2 would handle the throttles, monitor the power, and keep the flight log updated, and P3 would call out the speeds on take-off, monitor the aircraft systems, and oversee the operation.

The flight despatcher appeared on the flight deck less than 10min before departure with the details of weight distribution and aircraft trim, presented in the form of a load sheet. The tailplane incidence, or angle of the tailplane to the airflow, on the Trident was variable and could be adjusted to redress any out-of-balance forces in flight. Such action was known as trimming the aircraft. The tailplane angle required for balanced flight just after take-off was marked on the load sheet and would be checked and set accordingly on the tailplane indicator.

With the loading and trim details approved and all pre-flight checks completed, the engine start drill was commenced. Start clearance was requested and received on 121.7MHz at 15.39hrs GMT (all times GMT) 6min before departure.

Clearance R/T: 'Bealine 548 cleared to start.'

At the last minute, just as the engines were being started, the flight despatcher re-entered the flight deck and notified the captain that a Vanguard freighter crew were required in Brussels and would somehow have to be accommodated on the flight. The extra weight placed the aircraft over its payload carrying limit and obliged some load readjustment. The load alterations were undertaken and, a few minutes later with all engines running, the positioning crew boarded the aircraft. One was placed in the last remaining seat left vacant by a babe in arms, another was given a cabin crew seat, whilst the Vanguard captain, John Collins, was allocated the jump seat on the flight deck. Captain Collins had for some years been a Trident first officer before being transferred to the Vanguard fleet for his command, so was an experienced Trident pilot familiar with the aircraft's systems. He, however, would take no part in the duties of the flight. The three latecomers, added to the 109 passengers and six Trident crew — one stewardess, two stewards, and three pilots — brought the total on board to 118.

With the load adjustment complete, the load sheet updated and the trim rechecked, the aircraft was ready for departure. The passenger doors were closed at 15.58hrs. Papa India was parked by a passenger walkway facing nose in towards the terminal building and required a truck to push the aircraft clear. At 16.00hrs Captain Key selected 121.9MHz and called 'ground'.

Key R/T: 'Ground, Bealine 548 request push.'

With clearance received Captain Key informed the ground engineer on intercom that push back could be commenced. The push back truck by the nose revved its engines in response and Papa India moved away backwards from the gate. The ground engineer remained on intercom and walked out with the aircraft by the nose wheel. The aircraft was positioned on the taxiway and, with push back completed, the aircraft brakes were set to park. The Captain had a last word with the ground engineer confirming departure time of 16.00hrs and requested the all clear. The push back truck was driven away while the ground engineer disconnected his head-set and stood safely to one side, arm raised vertically indicating all clear.

Key R/T: 'Bealine 548, taxi.'

Ground R/T: 'Bealine 548, cleared to taxi to holding point two eight right.'

The time was 16.03hrs. While proceeding to the holding point the 'taxi drills' were commenced with Ticehurst reading over the intercom from the checklist to which the other two responded. Some aircraft types were known as 'shouting' aeroplanes where the crew conversed as normally as possible, leaving one ear free from the headset. On the Trident, however, all speech was conducted via the intercom system. The droops were selected out and the flaps set to 20°. During taxi Chief Steward Frederick Farey quickly visited the flight deck to confirm the cabin ready for take-off. Approaching the threshold the departure routeing was passed to Papa India at 16.06hrs.

Clearance R/T: 'Bealine 548, cleared to Brussels, Dover One departure, squawk standby six six one five.'

The clearance, as anticipated, was noted and read back by Ticehurst and was confirmed correct. 6615 was selected on the transponder in preparation for radar identification. On instruction, 118.2MHz was selected, and at 16.06:53 contact was established with the control tower.

Key R/T: 'Bealine 548, ready for take-off.'

Tower R/T: 'Bealine 548, cleared for take-off two eight right.'

As Papa India lined up on the runway a last minute hitch delayed their departure, and the tower was informed they had a small problem. This could very well have been the illumination of the stick push amber 'low pressure' warning light. The rain continued to fall heavily in the blustery wind. Just over half a minute later Key called for a second time to indicate readiness.

Tower R/T: 'Bealine 548, cleared for take-off.'

Key replied simply and abruptly, 548.

The time was now 16.08:24. In the wet conditions Key held the aircraft stationary on the toe brakes as Keighley advanced the throttles to the reduced power setting calculated by Ticehurst before departure. In spite of the maximum payload being carried, Papa India was still some 2,000kg short of maximum take-off weight because of the small fuel amount required for Brussels, and to reduce engine wear something less than full power could be set. As Keighley fine-tuned each thrust lever to the calculated setting on the P7 gauges, Key released the brakes and Papa India moved off quickly down the runway. Ticehurst checked the speed increasing while Key steered the aircraft using the nosewheel tiller. As the rudder became effective with speed Key transferred control, maintaing direction by guiding the rudder with the foot pedals. The wind was blowing from 210° at 17kt, weathercocking the aircraft nose into wind as the tail acted like a vane and the landing gear as a pivot. A good measure of right rudder was required to hold the aircraft on the centre line. Soon Ticehurst made the first call of 'one hundred knots'. Below this speed anyone observing danger could call stop, but after 100kt only the captain could call stop for an emergency, or the P2 for an obvious engine problem. Ticehurst's next call at 134kt was 'V1' the go or no go decision speed, followed quickly by 'rotate' at 139kt. Captain Key pulled back on the control column and rotated the aircraft to the required nose-up attitude for lift-off. Two seconds later at 145kt the wheels left the runway and Papa India became airborne. The take-off run had lasted 44sec and the time was now 16.09:14. Acceleration was still rapid and 'V2', the safe climb-out speed required in the event of losing an engine at V1, was quickly achieved and called at 152kt.

With the aircraft safely climbing the order 'undercarriage up' came from Key. Keighley immediately selected the lever to the up position and monitored the light sequence as the landing gear retracted into the wheel bays. Papa India was now buffeting wildly in the wet and turbulent conditions. As the wind began to bite, Key turned the aircraft slightly left to counteract drift and crabbed the aircraft along the extended runway centre line. The speed increased to 170kt and at a height of 355ft, 19sec into the air, the autopilot was engaged. Initial climb speed was based on the take-off safety speed of V2 plus 25kt and should have been 177kt indicated, but the speed lock was selected shortly after autopilot engagement with the speed still reading 170kt, 7kt lower than requirements.

Flying below the target speed in such turbulent conditions was definitely not to be recommended and clearly indicated Key's increasing discomfort and growing loss of concentration. As Papa India approached the marker, the beacon signal of medium pitched alternating dots and dashes could be heard on the headset accompanied by an amber light flashing in unison on each pilot's panel.

At 16.09:44, flying through intermittent cloud, passing 690ft, the left turn to track towards the Epsom radio beacon was initiated using the autopilot heading control knob. As the aircraft banked 20° to the left, Key made a last call to the tower.

Key R/T: 'Bealine 548, climbing as cleared.'

Tower R/T: '548, airborne at zero nine, contact one two eight decimal four. Good day.'

Papa India passed 1,000ft on the altimeter and entered thick cloud, buffeting and rocking in the turbulence. Key was slow to respond to the reply from the tower and not until 5sec after the call did he simply answer 'Roger'. Ticehurst quickly dialled 128.4MHz on the radio selector box.

In spite of the bumpy conditions, the autopilot was holding the speed quite well, with only a knot or two fluctuation on either side of 170kt. The time was now 16.10:00 and Keighley called 'ninety seconds', the noise abatement time. Three seconds later he selected the flaps from 20° to fully up. As the flaps were running up Keighley reduced power to the noise abatement power setting on the P7 gauge. With power reduction the autopilot decreased the climb angle in an attempt to maintain speed. At the same time Key called London on 128.4MHz.

Key R/T: 'Bealine 548 is climbing as cleared, passing 1,500ft.'

London R/T: '548, climb to flight level six zero.* Squawk six six one five.'

At this point the flaps fully retracted into the wing. Key's reply to the re-clearance from London was terse and non-standard.

Key R/T: 'Up to six zero.'

Ticehurst lifted his eyes to switch on the radar transponder, situated above the pilots' heads, which was set on stand-by with the code 6615 already selected, then made a note of the recleared height. Second Officer Keighley also clearly marked his flight log with this instruction to climb.

As the aircraft continued in a sustained 20° banked left turn towards Epsom, bumping around in the cloud, the autopilot equipment was coping less well and the speed dropped to 157kt, 20kt below target. A more alert pilot, checking progress, would certainly have intervened earlier to correct speed loss, but there is little doubt that by now Key was in pain with his attention wandering. A more confident co-pilot might just have detected the subtle incapacitation being

*At lower heights, the pressure altimeter is set to the local area pressure setting and indicates heights above mean sea level (MSL). Above a certain height, which varies throughout the world — at Heathrow at that time, for example, at heights of 6,000ft and above — the altimeter is set to a standard pressure setting which represents the MSL pressure on an average day. The height at which the altimeter is changed from local to standard pressure setting is known as the transition altitude. Heights below the transition altitude are expressed as an altitude in thousands of feet, eg 4,000ft, and above as a flight level in units of a hundred, eg flight level (FL) 70 is equivalent to 7,000ft, and FL 260 to 26,000ft.

experienced by his captain, but in the absence of obvious signs young Keighley could hardly be blamed for any inaction. He, or Ticehurst, may even have called 'speed' and been ignored.

At this normally quiet period of the flight, had the aircraft been climbing safely at target speed, the crew would have been more free to concentrate on the instrument departure route. Epsom and the minimum height required over the radio beacon of 3,000ft (the minimum height also required for droop retraction), were still a minute or two away. At Epsom the aircraft would be turned towards Dover. The engines would then be advanced to normal climb power and the speed increased. On reaching 225kt, the minimum droop retraction speed, the wings would be held level and the droops would be selected in. The speed would then be increased to the normal climb speed and the aircraft would continue on its way to destination.

After Papa India's noise abatement time, however, the turbulent conditions being experienced in cloud might not have been the only interruptions to have upset the crew in the quiet moments expected. At some time during the rough ride after take-off the three-way valve in the stick push ducting moved one-sixth out of position. It is quite possible that on power reduction the system pressure dropped below the lower limit, illuminating the 'low pressure' warning light. 'This light from its position might conceivably have been mistaken for the droop out of position warning light'. Turning in turbulence at such a low speed might also have triggered a stall warning, with the stick shaker operating momentarily. Ticehurst would have turned back to his instrument panel to check the integrity of the system, although under the circumstances it would almost certainly have been genuine. For a moment, at least, his attention would have been distracted. These events could have occurred simultaneously very soon after flap retraction and power reduction.

In Key's confused state, with knowledge of previous incidents, the alarm bells would have been ringing in his mind. Could he have assumed that young Keighley had retracted the droops instead of the flaps, and if so what would have been his reaction? Would he have called out a warning?

A shout of 'droops' might just have had the opposite effect than intended and might have prompted Keighley obediently to retract them. Such a call, however, would have undoubtedly been as brusque as Key's comments on the radio and would have alerted the others on the flight deck. If he genuinely believed the droops to have been retracted early, swift action would be necessary. In his disturbed state, alarmed by his confusion, he may have leaned across and quickly retracted the droops himself, thinking he was performing the exact opposite of his actions, ie re-extending them. As has been shown, mistakes of such magnitude have been made in the past by such experienced pilots as Key, without the influence of pain.

Perhaps, however, some confusion arose in Keighley's mind and prompted the young co-pilot to retract the droops. After noise abatement time, as Keighley was adjusting the throttles, Key was receiving his instruction for further climb to flight level six zero and was responding with his terse reply 'Up to six zero'. Immediately inboard of the droop lever, on P2's side, was situated the height acquire box. The altitude required was first set in the window and a knob pulled to arm the system. On reaching that altitude the autopilot would then capture the selected height. In this case the height acquire box would be set to the original instrument departure height restriction of 5,000ft, and it would normally be the task of the operating pilot to reselect the figures after clearance for further climb. To alter the height to a level above the transition altitude it would first be required to pull up the forward right-hand knob and to turn it to reset the millibar subscale to the standard setting of 1013. The same knob would then be pushed in and turned to reset 6000 in the window and the rear right hand knob pulled to arm the system. A few pilots of Key's vintage, however, were in the habit of requesting the P2 to set the height. Could Key have turned to his P2, just as he finished setting the power, and asked him to select the height? In his pain and discomfort it's likely that any such instruction would have been as abrupt as his radio call. Could he have pointed briefly to the area near the droop lever with the words 'Put it in' meaning the recleared altitude in the height acquire box. If such an instruction was followed by illumination of an amber light forward of the droop lever, the confusion, in this case, could very well be in Keighley's inexperienced mind. The correct command for retracting the droops is 'droops in', but if Keighley's mind was already thinking along these lines it is doubtful it he would have waited for a second brusque order. The droop lever can be moved easily and quickly, and with Ticehurst's attention momentarily drawn back to his panel, it is possible that Keighley selected the droops in.

Perhaps, instead, Key, with the 'Dublin incident' in mind, assumed that Keighley had selected the flaps down by mistake, rather than up, thus causing the low speed. His mind would picture flaps selected fully down with droops extended and his thought would be to have the flaps selected back to the up position. In such a configuration the flap and slat levers would sit side by side. Since the flaps were actually up, only one lever would be apparent and Key may have pointed directly at the droop lever with the words 'put that up'. The co-pilot may then have made the selection. With movement of the lever, Key's mind would picture the flaps set to the up position with the droops still extended, instead of the actual situation of both flaps and droops being selected up.

A similar situation can be envisaged if we imagine Key still relating his low speed to flap drag. He may simply have assumed that Keighley had forgotten to select the flaps up at noise abatement time. Once again a command of 'put that up' could have been actioned. Such theories, however, all seem to contain occurrences which one imagined would have drawn the attention of those on the flight deck to the situation, whereas no one seemed to know what had happened. Whoever moved the droop lever to up seems to have failed to appreciate the consequences of his action. Of course, the lack of response from the crew could simply have been the result of the overwhelming cacophony of noise which ensued, but is more likely to point to an insidious cause.

If Keighley had moved the droops, for example, he would have to be quick. The period between noise abatement time and droop-in selection was only 24sec. In these short moments Keighley had to select the flaps up, accurately reduce power in the rough conditions — not a simple task — and write up his log with the recleared height. Keighley, mistaking Key's command, could perhaps have rapidly moved the droop lever thus masking his action, as well as performing the other tasks, but he was not known for his swiftness. If Key was relating low speed to flap drag it would have been easier for him to take action himself. With the attention of the other two pilots drawn to their duties, Key, without saying anything, could simply have moved the only lever available, thinking he was retracting the flaps. If he had, it would not have been the first time such action had been taken, and would go some way to explaining the reason why the person who moved the droop lever failed to appreciate his actions. In Key's mind the flaps would now be retracted with the droops extended. If his hand was already in the area of the droop lever, setting the level in the height acquire box, he could have moved the droops easily without attracting attention.

It is possible that the height acquire box held the solution to the puzzle. If the recleared height of 6,000ft was set in the box then it is possible that Key selected the droops up in error at about the same time as setting the height. If 6,000ft was not entered, then it is possible that Keighley selected the droops up in error, incorrectly obeying the command 'put it in'.

The height acquire box was recovered from the wreckage but unfortunately was too badly damaged to be of use. In the end, none of the theories appear likely, and it is possible only to speculate. Since no cockpit voice recorder was on board it will never be known which pilot retracted the droops. At 16.10:24, however, only 6sec after Key's last sharp radio call and climbing through just 1,770ft, the droops were most certainly selected in. The speed had increased slightly to 162kt but, at 63kt below the minimum droop retraction speed, was dangerously low.

Papa India immediately entered the stall régime and the events which followed were swift and dramatic. Within one second of the droop lever movement, the flashing amber 'alert' lamp on each pilot's station operated, indicating a problem. The 'controls' window of the central instrument warning system (CIWS) display panel also illuminated outlining the fault area, and the droop 'out of position' lamp lit up specifying the cause. One second later the stick shaker operated, followed half a second later by the stick push. The amber 'stall recovery operate' lamps by the airspeed indicators illuminated with stick push operation, and, under certain circumstances, there may also have been a fleeting illumination of the red 'stall recovery fail' light situated nearby. The ram force pushing the control forward immediately disconnected the autopilot, illuminating the flashing red 'alert' lights at each pilot's station and also the red 'autopilot' window of the CIWS display panel. A loud audio autopilot disconnect warning was also transmitted to each pilot's headset. Clang! Clang! Clang! The above sequence would have occurred in something less than 3sec and would have caused the greatest distress, even to a healthy man. The effect on Key in his condition must have been overwhelming. It was probably at this point, with heart racing, that the pressure of the blood from the tiny ruptured vessel forced separation of the arterial lining. The pain could have been quite acute and his judgement would have been severely affected. In spite of his discomfort Key instinctively grasped the control column as the nose dropped violently. As the nose angle fell the stick push relented allowing Key to dive the aircraft and attempt to fly a recovery, but instead Papa India was held level and the wing bank taken off. No attempt was made to reselect the droops out and the crew were obviously at a loss as to their predicament.

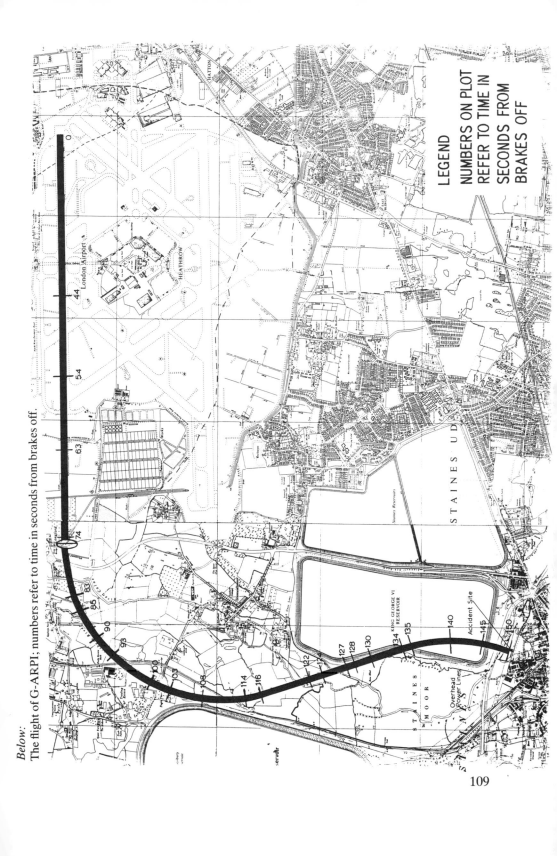

Below:
The flight of G-ARPI; numbers refer to time in seconds from brakes off.

LEGEND

NUMBERS ON PLOT
REFER TO TIME IN
SECONDS FROM
BRAKES OFF

109

At 16.10:32 the droops moved fully in at the wing leading edges. On retraction of the droops the centre of pressure, the point through which lift is assumed to act, moved forward. Since the variable incidence tailplane was positioned for aircraft trim with droops extended, a tail-heavy situation was immediately encountered. In effect, Papa India was trimmed into the stall. The nose pitched up again as the aircraft passed 1,560ft in a slight descent with the speed increasing to 177kt. The wings were still almost level with the aircraft heading due south. Two seconds after the movement of droops fully in, the stick push operated once more. Again the amber 'stall recover operate' lights illuminated. For a second time Key responded by holding the aircraft level rather than letting the nose dive towards the ground. In the turbulent cloud, with no visual reference and height low, the attempts of the stick push to plummet the aircraft earthwards must have been disturbing. Papa India, on the point of stall, was desperately short of speed and a steep descent to regain forward motion, or the reselection of droops, was essential. Had the nose been held firmly forward, full power applied, and the droops reselected (which would have taken about 10sec to run) Papa India might well have climbed away safely. In the confusion the cause of the problem was overlooked and no attempt was made to fly a recovery. The visual and aural warnings during these vital moments must have been overwhelming and could not have aided concentration. Key had also omitted to depress the cancel button on the control column for the autopilot disconnect warning, and the red flashing 'alert' lights and the loud clang, clang, clang audio warning continued throughout in their headsets.

A few moments after the second stick push the system pushed again for a third time. The nose dropped slightly below the horizon with the force but once more no full recovery was attempted. Suspicious of the integrity of the system the order was given by Key to dump the air supplying the stick pusher to which Ticehurst responded by selecting the lever to override. At 16.10:39 Papa India was still descending flatly from the sky, passing 1,275ft, with the aircraft banking 16° to the right. By now the airspeed had risen to 193kt, but Captain Key, perhaps sensing the aircraft had recovered and, probably with the normal climb speed of 177kt in mind, pulled back steadily on the control column at a constant indicated speed of 175kt. Papa India entered the 'true aerodynamic stall' and the nose pitched up once more. At 16.10:43 descending through 1,200ft, the aircraft locked into a deep stall with the nose angled 31° in the air. The indicated speed dropped to 54kt as the airspeed pointer settled on the lower stop. Recovery for Papa India was now impossible. At 1,000ft the aircraft broke cloud descending at a 60° angle at 4,500ft/min. It is doubtful, with heads down inside the flight deck, whether any of the crew would have seen the ground. In the driving rain, with the noise of the autopilot disconnect warning still clanging in the crew's ears, Papa India passed over the A30 highway and, just 36sec after the inadvertent droop lever movement, impacted on its belly near the Crooked Billet roundabout.

Walking in a field near Staines, 13-year old Trevor Burke, with his nine-year old brother Paul, witnessed the Trident plunge from the cloud.

'We were out with the dog and I looked up and saw the plane. It was just coming out of the mist when the engines stalled and it seemed it glided down. It was just like a dream. The plane just fell out of the sky.'

110

Above:
The crash site later in the evening. Lifting
equipment can be seen in the background.
Middlesex County Press

Right:
Aerial photograph showing open nature of
the crash site. Electric overhead cables
nearby remained untouched.
Middlesex County Press

Below:
The distribution of the wreckage.
The investigators' markings can be seen.
Middlesex County Press

'We just about saw it hit the ground,' said Trevor, 'because it was right in a clump of trees. When it did hit the ground the front bit hit first and the back bit was just blown away.'

The time was 16.11:00 GMT. The aircraft struck slightly nose down and the tail separated from the main fuselage. Papa India crashed about four miles from Heathrow on a narrow strip of waste ground between the A30 highway and the town of Staines, almost as if it had been deliberately guided to an open area. Not a single structure on the ground sustained as much as a scratch, and even power lines close by went untouched. The two boys witnessing the drama ran a quarter of a mile back home to summon the help of a nursing sister who lived near them. She was the first on the scene of the crash. A few motorists passing on the A30 also stopped to lend a hand. Surprisingly, the alarm to alert the rescue services was not raised by the controller monitoring his screen despite the fact that by then Papa India's radar bleep had vanished. The first emergency call was received from a man driving near Staines who had seen the aircraft crash in front of him. He stopped at a private house in London Road to phone the airport authorities.

A passing ambulance arrived at the crash site a few minutes later and within 15min of impact, police, fire brigade, and ambulances of the rescue services reached the scene, knocking down fences to get to the crash. As cutting tools were used to free those trapped inside a small fire broke out near the flight deck, but was quickly brought under control. The rescuers struggled on in the wind and rain with little thought for their own safety, but their efforts were to prove in vain.

As news of the disaster was announced on the radio, thousands of sightseers poured into the area, many of them families returning early in the rain from their Sunday drive. Police patrols were required around the crash site to discourage onlookers, and the roads in the accident area became clogged with vehicles, hampering movement of the rescue services. As it was, however, there was little need for haste. At the moment of impact the occupants of Papa India were beyond help. Only one person was pulled out alive, but deeply unconscious, from the wreckage and he died later on arrival at Ashford hospital. All 118 on board perished. The accident proved to be the worst in British aviation history and, but for the open site of the crash, could have been much worse.

The Paris DC-10 Crash 1974

At 19.19:48 local time on a cool summer's evening, American Airlines Flight 96 lifted off from runway zero three right (03R) at Detroit's Metropolitan Airport. The date was 11 June 1972. Flight 96 was a DC-10 (Series 10) scheduled service from Detroit to Buffalo and was lightly loaded with only 56 passengers and 11 crew. The aircraft had originally departed from Los Angeles and its final destination was La Guardia in New York City.

Detroit, Michigan, lies on the northwestern flank of Lake Erie while Buffalo, New York, lies at its most eastern point, with the US/Canada border running lengthwise along the centre of the lake. As the American Airlines DC-10 turned due east under 'departure' radar control the aircraft, registration N103AA, continued on its climb towards the Canadian town of Windsor in Ontario, with flaps and slats being selected up in sequence. The visibility was only 1½ miles in the still and hazy conditions and the cloud base was 4,500ft. Just under 3min after take-off the DC-10 entered cloud and at the same moment was cleared for further climb to flight level two one zero (21,000ft). First Officer (F/O) Peter Paige-Whitney was at the controls with the commander, Captain Bryce McCormick operating the radio. Second Officer (S/O) Clayton Burke occupied the flight engineer's station. Flight 96 was running about one hour late because of a number of delays, the most notable being the problem in closing the rear cargo door. Difficulties with the door had occurred on a number of occasions and it had taken 18min to close it in Los Angeles. In Detroit it had taken only 5min to shut the door but the ramp service agent had to use his knee to force down the locking handle. He was sufficiently concerned to mention it to an American Airlines mechanic who quickly checked the handle and found it safe. On the flight deck the 'door open' light extinguished confirming the door locked.

Departure control now instructed Flight 96 to contact Cleveland Centre on 126.4MHz. Although the short flight routed mostly through Canadian airspace, US controllers maintained contact. With the after take-off check complete and the aircraft climbing safely at 250kt, the first officer engaged the autopilot. Captain McCormick called Cleveland Centre climbing through 7,000ft and was instructed to select radar transponder code 1100. The DC-10 was further cleared to flight level 230. At 10,000ft the co-pilot set the vertical speed control to 1,000ft/min climb to reduce the climb angle for acceleration from 250kt to the normal cruise speed of 340kt. Passing about 11,500ft, directly over the town of Windsor with the speed increasing through 260kt, N103AA broke from cloud cover into a clear evening on top. A Boeing 747 could be seen high in the sky above.

Above:
Turkish Airlines took delivery of three DC-10-10s registered TC-JAU (illustrated here), TC-JAV and TC-JAY between December 1972 and February 1973. TC-JAV was the aircraft lost at Paris. *Alan J. Wright*

Below:
DC-10 cargo door.

'There goes a big one up there', remarked Captain McCormick.

The flight crew began to relax and to take in the view. Suddenly without warning, a resounding thud echoed from the rear of the aircraft. Someone yelled, 'Oh shit'. The rudder pedals 'just exploded' and smacked to the full left rudder position. The captain was resting his feet on the pedals and his right leg was thrown back against the seat with extreme force. F/O Paige-Whitney hit his head on the back of the seat. At the same moment the three thrust levers snapped back to flight idle with the number two (tail) engine throttle hitting the stop with a loud crack. A great rush of air swept past the flight crew throwing dirt and grit into their faces and blinding them with the dust. The stinging effect was like a fire cracker going off below their noses. The captain's headset was knocked from position and as the aircraft jerked the autopilot automatically disconnected. The captain's first thought was that the windshield had failed, but when his eyes cleared he could see it still in place. Disbelievingly he stretched his hand out to touch the window.

'What the hell was it, I wonder', called Captain McCormick.

One of the crew replied with a long whistle. The fire warning bell rang for number two engine and the cabin altitude warning horn sounded indicating the cabin air pressure had reduced to an altitude equivalent to over 10,000ft. The autopilot red 'disconnect' light flashed and a red failure flag appeared on the airspeed indicator. The captain then thought the radar dome might have gone. With the aircraft nose cover missing erratic airspeed indications could be expected. The first officer, by habit, still had his hands on the controls, but the aircraft banked slowly to the right out of control and the nose dropped sharply.

'We've hit something', said the second officer.

The co-pilot had been looking out and had not seen anything of another aircraft so thought it more likely to be disintegration of number two engine. That would explain the fire warning and the rudder problem although the fire indication subsequently proved to be false. Whatever the cause of the trouble there was no doubting the severity of the problem. In normal circumstances, with the cabin pressurisation warning horn sounding, the crew would have been tempted to commence an emergency descent, but the captain was reluctant to force the aircraft into a steep dive until the damage could be assessed. Flying at around 12,000ft few breathing problems would be experienced, although strictly speaking flight crew should wear oxygen masks above 10,000ft.

In the first few seconds after the loud bang the confusion in the cabin was just as great. At the rear of the aircraft Stewardess Sandra McConnell sat next to the right-hand emergency exit chatting to Stewardess Beatrice Copland who was positioned on the opposite side. Just forward of the girls was a small lounge bar normally used for cocktail service. Since there were very few passengers and the sector was short it was considered not worth offering the facility and no passengers were seated in the area. Both stewardesses had undone their seat belts and were thrown from their seats with the 'explosion'. Stewardess McConnell was thrown against the bar portion and landed by the edge of a gaping hole in the bar floor. She could feel herself slipping into it as the floor around crumbled into the gap. Stewardess Copland found herself lying in the hole and she could see into the cargo compartment. The bar unit was lying nearby and her head and one foot were trapped by the debris which had fallen from the ceiling. She called out for help.

Stewardess Carol McGhee usually worked in the cocktail lounge, but since her services were not required she sat strapped in by the front exit for longer than normal. Suddenly she heard a noise like a 'frump' and saw the escape hatch from the downstairs galley shoot up from the floor and strike a passenger on the head. The flight deck door burst open and out flew the crew's hats. A dusty rush of air gushed from the flight deck and the cabin filled with a 'cool fog'. Stewardess Carol Stevens was also still strapped in her seat when she heard a 'vroom' kind of noise. The cabin fogged and air rushed from front to rear of the aircraft. When the fog cleared she could see ceiling panels hanging down from the rear of the economy section and the bar lounge floor caved in. By the collapsed bar unit she could see Stewardess Copland pinned down the hole and could hear her cries for help. Stevens tried to call for support on the cabin interphone but without success, so ran forward to summon assistance from the other crew members.

In the service centre at the forward end of the aircraft, Head Stewardess Cydya Smith was already preparing coffee for the cabin service. Suddenly the galley lift door blew open and a 'smokey substance' billowed out. She was thrown off balance but somehow managed to hold herself upright. She also noticed ceiling panels drop and experienced a feeling of weightlessness. Immediately she thought of depressurisation and quickly checked to see if oxygen masks had dropped in the cabin, but they remained in position. She had no difficulty in breathing. At cabin altitudes above 10,000ft the warning horn on the flight deck sounds but it is only at cabin altitudes above 14,000ft that passenger oxygen masks drop automatically. She rushed forward to the flight deck, noticing on the way the captain's hat lying on the floor, and asked, 'Is everything all right up here?'.

'No', called the captain.

The co-pilot turned and shook his head. 'You go back to the cabin.'

Captain McCormick retrieved his headset from the back of the seat and quickly radioed Cleveland Centre declaring an emergency. There was little information he could give as they didn't know what had happened except that they had a serious problem. The co-pilot still handled the control column, but now relinquished flying to the captain who wanted to 'feel out' the controls himself.

'I think it's going to fly', assured the first officer.

The speed fell to 220kt as the DC-10 descended towards the cloud. If they could stay in the clear until the damage was assessed it would make life easier. The captain pulled back on the controls but soon ran out of elevator to arrest the descent. He pushed the thrust levers forward: the wing-mounted engines responded but number two tail engine remained in flight idle. The number two throttle could be moved backwards and forwards quite easily and it was obviously not attached to anything. The power from the number one and three (wing) engines effectively pitched the nose up and the aircraft managed to maintain 12,000ft. By good fortune McCormick had practised, in the simulator, flying the DC-10 on engine power only, assuming total loss of flying controls with hydraulic failure. The DC-10's engine positions permit relatively effective control when employed in this manner and Captain McCormick became quite adept at the task. He was now able to put his experimental endeavours to good use.

On returning to the cabin, Head Stewardess Smith comforted the passengers on the public address (PA) system and instructed them to stay in their seats, not to smoke, and to remain calm. As she finished speaking Stewardess Stevens raised the alarm that one of the girls was trapped at the back and required assistance. A male passenger offered to help and the three made their way rearwards, amazed at the destruction as they approached the tail.

With the aircraft now under control, the captain also lifted the PA handset to reassure the passengers. They had experienced a major problem, he explained, which was now resolved but would necessitate a return to Detroit. He would keep them informed of progress. The interphone from the cabin chimed on the flight deck. One of the stewardesses, unaware of the help being summoned for the trapped girl, called to say that Stewardess Copland required rescuing from a 'hole in the floor' and could someone lend a hand. Since the immediate flying problems seemed to be resolved, the flight engineer left his seat to help. Stewardess Copland managed on her own to wrench her head free from the debris trapping her in the gaping hole, then, leaving her shoe caught in the wreckage, pulled her foot out and was able to escape. She climbed out of the area, over a rear bulkhead, and on to one of the lounge seats where she was helped to her feet by the arriving rescuers. There was no sign of the other girl and Smith called out her name several times. As if from nowhere Stewardess McConnell appeared out of the debris and was helped into the cabin. In spite of the ordeal neither of the girls was hurt.

Second Officer Burke had only just left the flight deck and was still looking for his hat when the interphone sounded again. The two girls were now safe, he was informed, and his assistance was no longer required so he returned to his seat. In the cabin, passengers near the tail were moved forward from the extensively damaged rear section while others were comforted and nursed by the attendants. With the situation apparently under control Head Stewardess Smith returned to the flight deck to report to the captain. There were no serious injuries amongst the passengers and crew but there was a large hole in the floor and in the aft left-hand side of the fuselage. Captain McCormick instructed her to prepare for an emergency landing, and she returned to gather the other attendants in the service centre for a briefing.

Captain McCormick was still having control problems but was managing with difficulty to keep the stricken aircraft flying. The DC-10 could only bank about 15° in turns. F/O Paige-Whitney now operated the radio and requested emergency services for their arrival. Radar control was going to position the aircraft on a 20-mile final approach and cleared Flight 96 for descent. The best rate of descent that could be achieved was 200ft/min which was painfully slow. Radar control continued to vector the aircraft towards Detroit while the crew completed their checks. In the cabin the attendants prepared for an emergency landing and evacuation although it was not yet known if such an eventuality was likely. As a precaution, shoes and loose articles were collected, the brace position was demonstrated and the passengers were briefed on escape routes and the use of the slides. When all was ready one of the stewardesses went forward for confirmation that the escape slides would be used after landing. She popped her head through the flight deck door. 'Do you guys have a problem up here?' she asked.

In the very tense atmosphere, with the captain struggling to maintain control, the question seemed quite ridiculous and the crew roared with laughter. 'Yes, we have a problem', they called.

The captain said that he would activate the evacuation signal if necessary but they weren't sure yet what was going to happen. Captain McCormick then spoke to the passengers again, in as calm a voice as possible, apologising for the inconvenience and reassuring them that the aircraft was under control and that everything possible would be done for them on the return to Detroit.

The approach was begun at 20 miles out with 160kt indicated airspeed and a rate of descent of 600-700ft/min. Attempts at reducing speed resulted in unacceptable sink rates. In order to keep the aircraft aligned with the runway the entire approach was conducted with the nose pointing 5-10° to the right. The landing gear was successfully lowered and the remaining flaps extended just before touch-down. The DC-10 landed flat and fast 1,900ft in from the threshold. The landing was smooth, but once on the ground 'all hell broke loose'. Almost immediately Flight 96 veered off the runway and ploughed through the grass. The co-pilot took control of the reverse thrust levers, pulling maximum power on number one engine on the left and cancelling number three. This eased the aircraft left, back on to the hard tarmac of the runway, and the machine stopped about 1,000ft from the end. After the rough landing the captain felt an emergency evacuation might be prudent and activated the alarm. Fortunately, no further danger ensued and all the passengers quickly disembarked down the chutes. In the end only a few minor injuries were sustained. But for good fortune and the skill of the crew, the result could have been much worse.

The next morning National Transportation Safety Board (NTSB) accident investigators arrived on the scene and easily diagnosed the cause of the accident. Scuff marks on the rear cargo door's securing mechanism clearly showed that the door had not been properly locked, in spite of indications to the contrary. As Flight 96 had climbed out of Detroit, the cabin had been pressurised as usual to allow relatively normal breathing for the passengers and crew. The fuselage and doors are designed to prevent the pressurised air from bursting outwards into the rarefied atmosphere. Any failure of door, panel or window, however, results in air exhausting rapidly with force. In this case loads of up to five tons had been placed on the partially locked cargo door. Eventually the latches had sprung under the load and the door had blown open, causing explosive decompression of the aircraft. The door had been torn off by the airflow, damaging the left tailplane in the process. The pressurised cargo hold air had immediately exhausted to the atmosphere via the gaping hole and the cabin air pressure had placed an undue load on the floor. With insufficient venting in the cabin floor area the floor had simultaneously collapsed, trapping the bar unit in the open cargo door exit. Through the beams of the cabin floor ran control cables, hydraulic pipes and wiring of which a number had been severed or jammed, resulting in the control difficulties. Fortunately sufficient control systems had remained intact to sustain stable flight. It had been a close-run event and swift and decisive action would be required to prevent a recurrence.

The 'Windsor incident', as the near disaster became known, prompted an inquiry by American Airlines, the NTSB, the Federal Aviation Administration

(FAA) and McDonnell Douglas into the locking difficulties of the rear cargo door. A division of General Dynamics, Convair, the manufacturer of the door, was also involved. The FAA's Western Regional Office, one of 11 such establishments in the US, was responsible for regulating California's aviation industry. Since the state contained the biggest concentration of aviation related manufacturing talent in the world, the regional office's administration duties were a formidable undertaking. The two great rivals, Douglas and Lockheed, both resided within its jurisdiction. Head of the FAA's Western Regional office was Arvin Basnight, a career public servant whose ability was equal to the task. Shortly after the Windsor incident, on 13 June, Dick Sliff, Basnight's head of aircraft engineering, contacted Douglas at Long Beach regarding the cargo door problems. The company's attitude was less than helpful, and it was only after some agitation that information was made available. The documents examined by Sliff made interesting reading and revealed that there had been about 100 previous reports of difficulties in closing the door. Some even more damning documents, however, were not revealed. In particular, one written by Dan Applegate, an engineer of Convair, the cargo door designers, expressed extreme concern with safety aspects of the door. Had Sliff reviewed this information the story might have been different. In the normal course of events, records received by the manufacturing company from airlines are passed to the FAA for correlation in the form of Maintenance Reliability Reports. In this case, McDonnell Douglas, ever wary of adverse publicity for an aircraft in a highly competitive market, had not honoured the arrangement (although there was no legal requirement for them to do so) and the FAA were quite unaware of the troubles. Before the Windsor incident, however, the company had already attempted to take matters in hand and had issued Service Bulletins to the four airlines operating DC-10s (American, Continental, National and United) recommending rewiring of the door's electric actuators. It was hoped that by increasing the power of the actuators with the use of heavier gauge wire the latches would be driven more fully home. Rewiring of DC-10 cargo doors was still in progress at the time of the incident but Captain McCormick's aircraft had not been modified.

To those involved at the local FAA office the proposed remedy did not seem adequate. If the DC-10 was to continue flying over the busy summer which lay ahead some interim measure was required until a more effective solution could be devised. The problem was that the actuators were not driving the latch linkage to the over-centre position (see diagram) and it was frequently necessary to shut the doors manually using a hand crank. With the linkages correctly placed, closure of the outside door handle slid a lock pin in place to secure the system. With over-centre not achieved a restraining flange prevented engagement of the locking pin thus interrupting closure of the door handle. Before Flight 96 had departed from Detroit, the door had been only partially secured and the locking pin had jammed on the restraining flange. The ramp service agent had then forced the handle home with his knee, distorting the locking pin rods in the process. This had given the false impression that the door had not only closed but had locked, in spite of the fact that the locking pins were not in position. In flight, as the interior pressurised, the enormous force on the latches had been transmitted via the

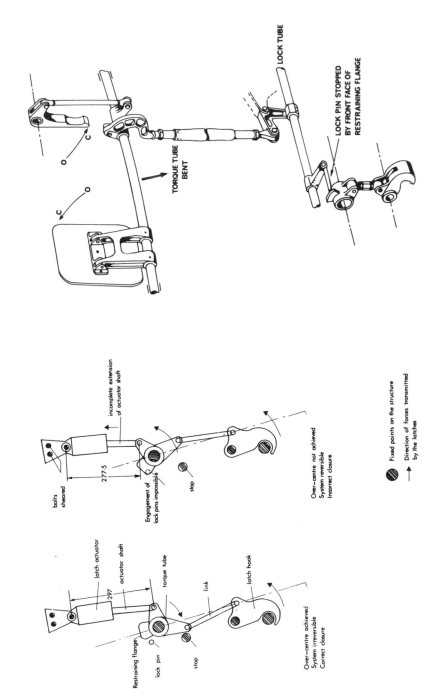

Above:
The problem of forced closure.

Above:
DC-10 aft cargo door latch closing system.

incorrectly positioned latching mechanism to the actuator bolts which had sheared under the strain with the inevitable results.

One simple solution seemed to be provision of a one-inch diameter peep-hole of toughened glass through which a locking pin could be checked in position. It would, of course, require a certain amount of dedication from the individual closing the door to check properly the pin's position, especially on a cold, wet and windy night. It was not the complete answer, but something immediate was necessary to help prevent a recurrence of the incident over Windsor. The peep-hole requirement would also be backed by the full force of federal law to ensure the airline's compliance. The document employed to enforce such an order is known as an Airworthiness Directive (AD) and is issued only in matters concerning safety. The release of the AD, unfortunately for McDonnell Douglas, would make public the circumstances, and being an airworthiness requirement the costs of the modification would have to be borne by the manufacturer. Early on the morning of 16 June, the draft of the AD was telexed to the FAA headquarters in Washington for approval, but it soon became apparent that consent was not to be forthcoming. Before 09.00hrs, Basnight received a telephone call from Jackson McGowan, the president of the Douglas Division of McDonnell Douglas. The previous evening McGowan had been in conversation on the phone with Jack Shaffer, the head of the FAA in Washington DC, and was now relaying his information to the Western Regional Office. After some discussion, McGowan said, he and Shaffer had agreed that 'the corrective measures could be undertaken as a product of a gentlemen's agreement, thereby not requiring the issuance of an FAA Airworthiness Directive'. The staff at the local FAA office was somewhat taken aback by this turn of events, especially receiving the details from Douglas who had been less than helpful. Much telephoning ensued between California and Washington DC, but it soon became clear that an AD was not to be issued and that Basnight and his staff were being by-passed in discussion with the airlines. A telephone conference was arranged between Douglas, the FAA in Washington DC, and the four airlines to agree on proposals for modifying the cargo door. Agreement was reached to continue with the wiring programme and to placard the doors with a warning for the ground staff not to use a force in excess of 50lb when closing the door handle. How this figure was to be gauged was not explained. It was a totally inadequate response to a very dangerous situation.

The FAA plays contradictory roles of somewhat incongruous dimensions being both watchdog and promoter of aviation in the USA. In the America of President Richard Nixon, during the years leading to the disgrace at Watergate, little doubt was left to the government's attitude. The advancement of trade, commerce and industry was of prime importance, and the small affair of a defective cargo door was not going to be allowed to rock the boat, or for that matter, the DC-10. McDonnell Douglas had some very powerful friends in high places. The federal organisations are usually headed by political appointees nominated by the White House, and the FAA then and now is no exception. Nixon's intentions, however, of weakening the independence of the federal agencies by creating organisations subservient to his personal command were being effectively implemented and were beginning to take their toll. The head of the FAA, John Shaffer, a man of little commercial experience, was a political appointee. The decline suffered by the FAA under Shaffer, a basically honest man but overprotective of the industry, was far-reaching and took many years to resolve. The board of the NTSB consisted of five men all of whom were political appointees. Working under the NTSB chairman, John Reed, was Ernest Weiss, a very able administrator but known to be an active Democrat. Plans were laid to oust Weiss from his position. Head of the NTSB's Bureau of Air Safety was another astute gentleman, Charles (Chuck) Miller, a renowned safety expert and distinguished engineer. His contribution to air safety was impressive. In the early days of the Boeing 747, turbine blade failures resulting in engine break-downs were becoming embarrassing with some aircraft losing more than one engine on the same flight, but little was being done by the FAA to solve the problem. In October 1970, Miller, frustrated by the FAA's inaction, vented his impatience by publically criticising the Authority. He pressed for measures to resolve the 'potentially catastrophic' situation. By the end of the year the FAA relented and issued an AD against the problem engines, but only one requiring regular inspection of the turbine blades. It was several months before the blade manufacturers resolved the situation, by which time another 16 incidents had occurred. The fact that the potential disaster predicted by Miller had not materialised only served to encourage the FAA in their policy of seeking voluntary responses. By such practices it was intended to resolve aviation's problems quietly and to maintain the good image of the industry. But would it be enough?

In March 1971, under pressure from the White House, the NTSB chairman John Reed demoted Weiss and replaced him with another political appointee, Richard Spears, whose qualifications were only the minimum for the job. It was inevitable that Spears and Miller, now a thorn deep in the FAA's side, would clash, and by the summer of 1972, at the height of the DC-10 cargo door problems, the hostility between the two men broke into the open. At a time when Miller's expertise was most needed, his attentions were diverted by systematic interference in the affairs of the Bureau of Aviation Safety, and in particular Spears's campaign against the Accident Prevention Branch. Miller was not going to be allowed publicly to criticise the FAA and get away with it. *

On 6 July 1972 the NTSB formally presented its recommendations to the FAA: the door should be rendered 'physically impossible' to close incorrectly, modifications should be made to the floor and vents, and the floor strength should be increased. The response at the time from the FAA and McDonnell Douglas was generally to ignore the proposals. Service Bulletin (SB) 52-27 regarding the wiring improvements was followed over July and August by two more notices. SB 52-35 recommended the installation of the peep-hole with a diagram by the door

*By April 1973 Miller found himself facing Spears's allegations of incompetence which were lodged with the NTSB, and in August the Board sat in judgement. The verdict was improve or be fired. Miller counter-attacked by taking the matter to a higher authority and later an inquiry was convened by the Senate Commerce Committee. Before it could be resolved, however, Miller became ill with heart problems and a long stretch of periodic sick leave ensued. In December 1974 he took early retirement from the NTSB on the grounds of ill health. Fortunately, he fully recovered and took up lecturing in air safety.

frame indicating the safe and unsafe positions. The blue paper on which the bulletin was raised indicated its relevance to safety. SB 52-37, routinely printed on white paper, recommended the installation of a support plate to prevent distortion of the locking pin rods, and also extension of the locking pin travel by one-quarter of an inch to make jamming of the pins more obvious. Douglas's response, on the surface, seemed to be reasonable, and the proposed modifications at least adequate for the immediate future. Had the recommendations been fully implemented the door locking problems may have been solved, but the lack of urgency with which they were delivered left a lot to be desired. Without an AD to enforce the issue the four airlines, all US, flying the 39 DC-10s in service at the time, were slow to apply the modifications. Three months later, by October 1972, only five of the total fleet had been modified. By the end of the year 18 had still not been altered, and one was still operating without a support plate in 1974.

In the summer of 1972, at the height of the DC-10 cargo door controversy, Lockheed with its TriStar, and McDonnell Douglas with its two versions of the DC-10 (Series 10 and Series 30), were each looking overseas for further orders. Both were simultaneously wooing Turkish Airlines who seemed a good prospect for sales, and the outcome could influence deals with other airlines. In the end McDonnell Douglas won the order and three DC-10-10 aircraft were duly delivered to Turkey in December 1972. In the meantime Nixon's first full term of office came to an end and he was successfully re-elected. In the course of such events tradition demands that all senior political appointees offer their resignation and the head of the FAA, John Shaffer, acted accordingly. No one was more surprised than Shaffer when his resignation was accepted! The following year he was replaced by Alexander Butterfield who had formerly been involved with internal security at the White House. Before Butterfield's arrival, however, and perhaps influenced by Shaffer's imminent departure, the FAA began reviewing their attitude regarding possible floor damage on wide bodied jets as a result of explosive decompression. The Dutch equivalent of the FAA, the RLD, had been involved since the Royal Dutch Airlines (KLM), decision to buy the DC-10, and had been expressing concern for some time. In September 1972, RLD representatives met with the FAA and McDonnell Douglas to discuss the issue. Douglas was quick to point out that the DC-10 floor strength met FAA requirements, but, the Dutch countered, in the light of the Windsor incident the regulations must be considered inadequate. This view was supported by NTSB research. The outcome of the talks was inconclusive with the FAA supporting McDonnell Douglas and agreeing to differ with RLD.

By February 1973 the FAA was beginning to admit the error of its ways and was now urging the big three aircraft manufacturers to consider strengthening their floors, increasing venting, and re-routeing essential control lines away from the floor. Both Boeing and Lockheed, perhaps understandably, were indignant at being drawn into the argument surrounding another maker's aircraft. McDonnell Douglas insisted that present regulations were satisfactory. In June 1973 the FAA asked its regional officers to obtain technical details about big jet floors, but with surprising unwillingness on the part of the Western Regional Office, it was not until February of 1974 that the request was followed through. The reaction of all three manufacturers was predictable as each baulked at the thought of the expense involved in a detailed study. The Dutch had by now reluctantly certificated the DC-10, and the FAA's rather belated attempts at action were not well received. On 25 February 1974, McDonnell Douglas replied stating that if the FAA insisted on a study then the government should bear the cost.

Less than a week later, on 2 March, the English and French rugby football teams faced each other in the Parc de Prince in Paris for what proved to be a fast and hard fought game. The result was a 12-12 draw. The estimated 30,000 English fans in attendance swelled the visitors to the city and placed an extra strain on the transport facilities between Paris and London. To make matters worse British European Airways (BEA) ground engineers at Heathrow had called a short strike in support of a pay claim against the company and all their European air services were grounded. On the day following the rugby match, Sunday 3 March, the chaotic situation at Paris's Orly Airport was compounded by other travellers arriving from elsewhere in Europe, placing themselves one step nearer home in an attempt to return to London during the weekend. The result was a gigantic headache for BEA staff at Orly as they tried desperately to find seats on other airlines to get their passengers to London.

Turkish Airlines (Turk Hava Yollari — THY) Flight 981 from Turkey to Britain that Sunday was an ideal contender for the beleaguered BEA staff. The aircraft was a DC-10 (Series 10) of 345-seat capacity, routeing Ankara-Istanbul-Paris-London, and in excess of 200 seats were calculated as being available on the final Paris-London sector. The Turkish DC-10, registration TC-JAV, landed on schedule at just after 11.00hrs local time (10.02hrs GMT) with 168 passengers on board. The aircraft parked at stand A2 at the west satellite of the Orly-Sud air terminal and 50 passengers disembarked while those in transit remained on board for the one hour stop-over. Scheduled departure was local mid-day. Turkish Airlines had a few staff on hand to supervise operations in Paris but much of the ramp area work was subcontracted to Samor Co. Aircraft loading was one of Samor's responsibilities and its personnel were instructed in DC-10 cargo door closing techniques. After switching on the power a button in a recessed control panel adjacent to the door frame was to be pressed to power the actuators. With the door shut the operator was to continue depressing the button for a further 10sec to ensure correct positioning of the latches. The external handle was then to be placed flush with the door to engage the locking pins and to close a small vent door (see diagram). This was stated as being the indication that the door was safe. A warning was also issued on the use of any force in closing the door handle. A final check of the locking pin position through the peep-hole was not part of the duties of Samor's staff but was the responsibility of Turkish Airlines. The resident THY ground engineer at Paris, Osman Zeytin, was in Istanbul on a course, and in his absence Flight 981's transit was supervised by another ground engineer, Engin Ucok, who was to continue with the flight to London.

To facilitate procedures at Paris, THY had distributed the baggage and mail conveniently throughout the cargo holds. All the load intended for London had been placed in the forward cargo compartment which was not opened. Passengers boarding at Orly had their bags placed in the central compartment while the aft cargo compartment contained only baggage and mail destined for Paris. It was

124

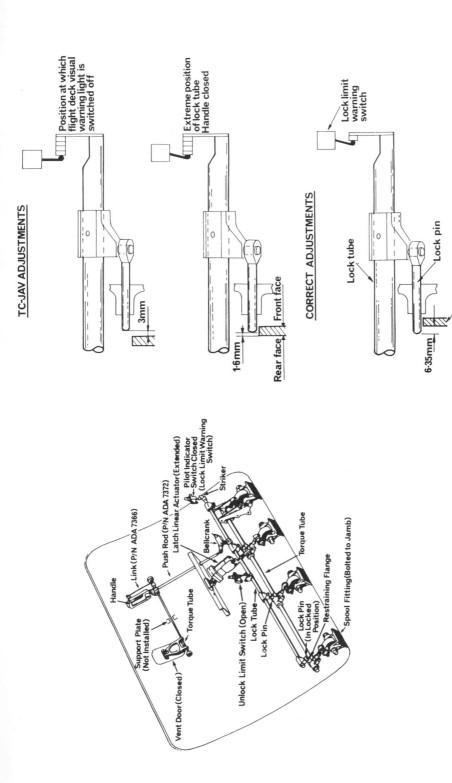

TC-JAV ADJUSTMENTS

Position at which flight deck visual warning light is switched off

3mm

Extreme position of lock tube Handle closed

1·6mm

Front face

Rear face

CORRECT ADJUSTMENTS

Lock limit warning switch

Lock tube

Lock pin

Restraining flange of latch crank

6·35mm

Above:
The faulty adjustments showing how the warning light in the cockpit is actuated.

Handle

Link (P/N ADA 7366)

Push Rod (P/N ADA 7372)

Latch Linear Actuator (Extended)

Pilot Indicator Switch Closed (Lock Limit Warning Switch)

Striker

Bellcrank

Torque Tube

Support Plate (Not Installed)

Torque Tube

Unlock Limit Switch (Open)

Lock Tube

Lock Pin

Lock Pin (In Locked Position)

Restraining Flange

Spool Fitting (Bolted to Jamb)

Vent Door (Closed)

Above:
Closing and locking mechanism.

completely emptied. Nothing was loaded in the rear hold and the door was closed at 10.35hrs GMT (all times in GMT) by one of Samor's staff, Mahommed Mahmoudi. He was a 39-year-old Algerian expatriate who could speak two languages fluently, Arabic and French, and could read and write in them both. He could not read English, however, so could not understand the placards by the door indicating the safe and unsafe positions of the locking pins. He had on a number of occasions seen Zeytin, the THY resident ground engineer, place his eye at the peep-hole but he was not aware of its function. Mahmoudi had correctly followed the door closing procedure and nothing had given him cause for concern. THY had experienced numerous difficuties with the opening and closing of TC-JAV's rear cargo door and a number of times it had to be closed using the hand crank, but on this occasion it shut without any undue effort. The final check of the safe condition of the door should have been conducted by one of the THY staff, but neither Zeytin's replacement, the ground engineer Ucok, nor the DC-10's flight engineer, Erhan Ozer, checked the peep-hole. Had they done so they would have seen that the locking pins were not in place. The latches had not been powered home fully and the latch linkage had not been driven to the safe over-centre position. As in the Windsor incident all indications were that the door was closed and locked. History was about to repeat itself. But in the Windsor incident the handle had been forced in position with a knee while Mahmoudi in Paris had placed the handle flush with ease. In fact, it had been almost too easy.

In the Douglas factory at Long Beach, THY's DC-10, TC-JAV, was designated the code Ship 29 during it construction. It was completed in the summer of 1972 and delivered to THY at the end of the year. The paper work for Ship 29 clearly stated that the current Service Bulletins had been implemented but, because of an oversight, the requirements of SB 52-37 (extension of the locking pin travel and fitting of the locking pin rod support plate) had not been done. After delivery to THY, an adjustment to the locking pin travel was made but the aircraft was still awaiting fitment of the support plate. By a gross blunder, however, alteration of the locking pin travel was incorrectly applied. Instead of the travel being extended to seat the pin properly when locking the door, or to make obvious the jamming of the door handle with improper setting of the latch linkage, the locking pin travel was actually *decreased*. Even in the locked position the pins were hardly effective (see diagram).

McDonnell Douglas had calculated that with the locking pin travel extended as required, a force of 215lb was needed to close the handle with the latch linkage incorrectly set, and with the support plate also fitted a force of 430lb (beyond human strength) was necessary. With the locking pin travel decreased as on TC-JAV, a force of only 13lb was required to place the door handle flush with the door still unlocked. The incorrect adjustment also affected the lock limit warning switch which illuminated a 'door open' light on the flight deck to alert crews of incorrect positioning of the lock pins. The mis-rigging resulted in the lock limit microswitch failing to extinguish the flight deck 'door open' warning light, even when the door was locked. At some stage the lock limit switch striker had been extended by the addition of extra shims to permit a more positive contact (see diagram). The result of the tampering was that the flight deck 'door open' light now extinguished even when the door was still unlocked.

126

To all intents and purposes, therefore, TC-JAV's rear cargo door was closed and locked. Only a visual inspection through the one-inch peep-hole would have indicated the true unsafe condition, and in the absence of the resident ground engineer that was overlooked.

From the chaos of the terminal building passengers began to board Flight 981 and take their seats. In all, 216 people were boarded to join those waiting on the aircraft giving a passenger figure of 334. In the confusion 10 or so seats remained unallocated. The crew of 11 (three flight crew and eight cabin staff), plus the ground engineer travelling on board, brought the grand total to 346. The boarding of so many extra passengers in such difficult circumstances resulted in the inevitable delay, and it was not until 11.11hrs GMT that THY first contacted Orly 'pre-flight' on 120.5MHz for departure instructions. Flight 981 was assigned departure route 18 from runway 08 — routeing via Tournan intersection, Coulommiers and Montdidier, with an initial climb to flight level 40 — and was instructed to select radar transponder code 2355 on take-off. Departure route 18 tracked aircraft east and then north to avoid overflying Paris. THY then changed to 'ground' frequency 121.7MHz at 11.14hrs and after completion of boarding and engine start was cleared at 11.24hrs to taxi to runway 08. The conditions were fine

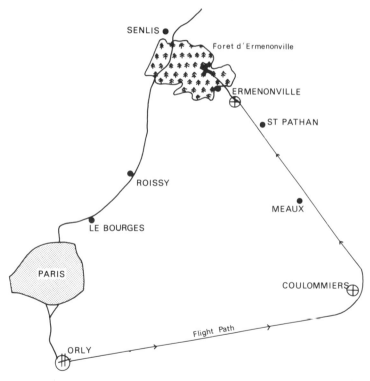

Above:
DC-10 flight path.

with a light wind, some patchy cloud, a temperature of 6°C and good visibility. On the flight deck Captain Nejat Berkoz and his crew of F/O Oral Ulusman and F/E Erhan Ozer completed the before take-off checks and, approaching the runway, changed to 'tower' on 118.7MHz. Flight 981 was cleared to line up and take-off, and at 11.30:30 the DC-10 weighing 163 tonnes lifted off from the runway. One and a half minutes later 'departure' was contacted on 127.75MHz and clearance was received for further climb to flight level 60. The after take-off check was completed and the seat belt sign switched off, but most passengers preferred to remain seated with their lap straps fastened. The autopilot was engaged. TC-JAV reported level at 60 and was instructed to contact Paris (North) on 131.35MHz. On contact with the area controller, Flight 981 was cleared at 11.36hrs for further climb to flight level 230 and instructed to turn left to Montdidier. Five routine communications passed between Flight 981 and Paris (North) in the course of the turn and at 11.38hrs the DC-10 stabilised on a heading of 346°, climbing through flight level 90 at a speed of 300kt. The aircraft interior continued to pressurise as the DC-10 climbed into the rarefied atmosphere, placing a force of almost five tons on the unlocked rear cargo door. At 11.39:56 Flight 981 passed 11,500ft climbing over the village of Saint-Pathus at a rate of climb of 2,200ft/min and a speed of 300kt. Suddenly the door latching mechanism could no longer withstand the strain and the door burst open and tore from the side of the fuselage. The heavily laden rear section of the cabin floor collapsed completely with the force of the explosive decompression and the last two rows of triple seat units above the door, with six passengers and parts of the aircraft, were ejected. The cabin fogged and dust swirled in the rush of air.

On the flight deck the crew were taken completely by surprise. The throttles snapped closed and the autopilot disconnected. Almost immediately the DC-10 banked to the left and the nose pitched down rapidly.

'Oops. Aw, aw,' someone exclaimed.

The co-pilot grabbed the controls as Flight 981 dived towards the ground and the cabin pressurisation warning horn sounded.

11.40:05, Captain Berkoz: 'What happened?'

F/O Ulusman: 'The fuselage has burst.'

11.40:07, Captain Berkoz: 'Are you sure?'

The control cables and hydraulic lines running rearwards below the floor were torn from their tracks, or jammed, and elevator and stabiliser controls were lost. The rudder seized at an angle of 10° to the left. There was now no means by which the crew, in spite of their efforts, could regain sufficient control of the aircraft. As the nose dropped the speed began to build up.

11.40:12, Captain Berkoz: 'Bring it up, pull her nose up.'

At this moment, 11.40:13, a radio transmission was received from Flight 981, either by chance or intention, by the radar controller at Paris (North). He could hear a heavy background noise with Turkish being spoken and the noise of the cabin pressurisation warning horn sounding. As the crew fought to regain control the transmission continued.

F/O Ulusman: 'I can't bring it up — she doesn't respond.'

The aircraft continued to accelerate towards the ground. By 11.40:18 the nose was pitched down 20° with the speed increasing through 362kt.

11.40:19, F/E Ozer: 'Nothing is left.'

At 11.40:21, with the DC-10 descending into the more dense atmosphere, the pressurisation warning horn ceased. Passing 7,200ft the speed increased to 400kt, and the aircraft continued banking in a left turn as it raced earthwards.

F/O Ulusman: 'Seven thousand feet.'

A second or so later the overspeed warning sounded as the speed edged beyond the 'never exceed' speed. At the same moment the Paris (North) controller noticed TC-JAV's flight label disappear from the secondary radar scope. On the primary radar screen a thin sliver of light representing the ejected parts detached itself from the echo and remained stationary while the DC-10's trace curved to the west.

11.40:28, Captain Berkoz: 'Hydraulics?'

F/O Ulusman: 'We have lost it . . . oops, oops.'

At 11.40:31 the nose down pitch began progressively to decrease and the speed stabilised at 430kt. The radio transmission being received on the ground abruptly ended at 11.40:41 as the controller continued to monitor the DC-10's progress.

11.40:50, Captain Berkoz: 'It looks like we are going to hit the ground.'

11.40:52, Captain Berkoz: 'Speed.'

The overspeed warning continued to sound throughout, but fainter than at the beginning.

11.40:57, Captain Berkoz: 'Oops.'

The DC-10's angle of descent stabilised at a shallow angle but the machine continued at great speed towards the ground. TC-JAV was beyond help. At 11.41:04 and 11.41:06 further short radio transmissions were received by the ground controller from the DC-10. These were the last to be heard. Flight 981 struck the trees of the Ermenonville forest at 11.41:08 at a speed of 430kt in almost level flight, but with the wings banked 17° to the left. At 11.41:31 the DC-10 impacted in a rugged valley at a place known as Bosquet De Dammartin, 37km (20 miles) northeast of Paris, only 77sec after the door burst open. The aircraft 'cut a swath through the forest some 700m long by 100m wide', and literally disintegrated as it struck. All 346 on board perished. With the force of the impact TC-JAV exploded into a million tiny pieces and scattered over a wide area. There was virtually no fire. Nothing had a chance to burn.

In the control centre the radar controller watched helplessly as the echo disappeared from his screen. Many times he tried in vain to contact Flight 981 and with no reply he raised the alarm. Soon a large-scale rescue operation was put into action and the first rescuers arrived quicky on the scene at 12.15hrs. There was little they could do except start to clean up the mess. Experts who witnessed the devastation stated that they had never seen an aircraft disintegrate so completely over such a wide area.

Within 22min of the disaster, even before the first rescuers arrived at the site, the BBC broadcast news of the crash. It was the accident the world had feared: the first involving a fully laden wide body jet since their introduction into service only four years earlier. The worst previous had killed 176 when a Boeing 707 crashed on landing at Kano, Nigeria, returning with Muslims from a pilgrimage to Mecca in January 1973. The THY DC-10 catastrophe, with all 346 on board killed, was at the time the worst disaster in civil aviation's history. It seemed a staggering number to perish in one air crash. Only hours after the accident journalists were on the scene to report the details to the world. As with the Trident crash at Staines, the spreading of the news attracted thousands of the morbidly curious in seach of a glimpse of the carnage. Police were required to keep the sightseers at bay.

Immediately FAA and NTSB investigators, including Chuck Miller, still head of the Aviation Safety Bureau in spite of his trials and ill health, arrived in France from the US to examine the wreckage. Inspection of the door soon revealed the cause of the tragedy. The next day, 7 March, the FAA at last issued an Airworthiness Directive requiring compulsory application of the modifications outlined in the previous Service Bulletins. Butterfield, the head of the FAA, to his credit ordered an immediate investigation into his own organisation's handling of the DC-10 cargo door affair. In the process of the inquiry a survey of the records of problems experienced with the cargo door between October 1973 and March 1974 revealed 1,000 separate incidents. The report was submitted on 19 April 1974 and was highly critical of the cargo door design and the system which permitted its certification. A special sub-committee of the House of Representatives also reviewed the affair and determined that between the Windsor incident of June 1972 and the Paris crash of 1974, 'through regulator nonfeasances, thousands of lives were unjustifiably put at risk'. Demands by the FAA for wide body floor strengthening were followed in July 1975 with the introduction of legislation requiring all wide body floors to withstand the decompressive effect of a 20sq ft (the area of the DC-10 rear cargo door is a little over 14.5sq ft) hole appearing in the fuselage.

The report of the official inquiry clearly outlined the imperfections of the cargo door latching mechanism and the risks entailed in its faulty operation. 'All these risks', the report concluded, 'had already become evident 19 months earlier, at the time of the Windsor incident, but no efficacious corrective action had followed'.

The BEA/Inex-Adria Mid-air Collision   1976

The town of Split lies on the coast of Yugoslavia almost exactly between Italy to the north and Albania to the south. Since earliest times travellers have been attracted to its sunny shores and the warm Adriatic Sea. In the third century AD the Roman Emperor Diocles built a magnificent fortified palace for his retirement on an attractive cove which now forms the harbour of the modern town. Many parts of the building can still be seen to this day. In the Emperor's footsteps have followed armies of tourists from the cold north, seeking sun and relaxation in the pleasant resort. West German holidaymakers, especially, have found it a popular destination.

On the morning of 10 September 1976, a party from Cologne holidaying in the Split area prepared for their return home after an enjoyable stay. As they faced the trauma of check-in at busy Split Airport, another group of travellers was already beginning its journey from the distant airport of London, Heathrow.

British Airways scheduled flight BE476 from London to Istanbul lifted off from Heathrow at 08.32hrs GMT (09.32hrs local). The Trident 3B, registration G-AWZT (Zulu Tango), was less than half full for the 3½hr journey to Turkey, with only 54 passengers on board. The travellers formed a disparate group of various nationalities and even included one stateless person. Captain Dennis Tann commanded the Trident with First Officer (F/O) Brian Helm as co-pilot and F/O Martin Flint in the P3 position. Six cabin crew tended the passengers' needs. Flight BE476 crossed the Channel at Dover, then proceeded southeast over Belgium, passing overhead Brussells, and continued to West Germany. About 1¼hr after take-off, with breakfast completed, Flight BE476 passed over the city of Munich and reported its position at 09.43hrs. The Trident settled on airway Upper Blue 1 (UB1), flying at flight level 330 towards the radio beacon of Villach in the south of Austria. At such quiet moments in the cruise, with eyes and ears still alert, crews often keep their minds active with the daily cross-word puzzle. F/O Helm was struggling with a clue and engaged the assistance of the others. As the quiet discussion ensued Trident Zulu Tango slipped from West Germany into Austrian airspace at 09.48hrs. At precisely the same moment, Flight JP550, a DC-9 of the Yugoslav charter airline Inex-Adria Aviopromet, lifted off from Split airport carrying the party of tourists returning to Cologne. The DC-9, registration YU-AJR (Juliet Romeo), was almost full with 108 passengers, of whom all, except one Yugoslav, were West Germans. On the DC-9's flight deck were two pilots, Captain Joze Krumpak and F/O Dusan Ivanus. In the cabin the passengers were attended by only three stewardesses who had their work cut out catering for so many people on the short 2½-hour journey.

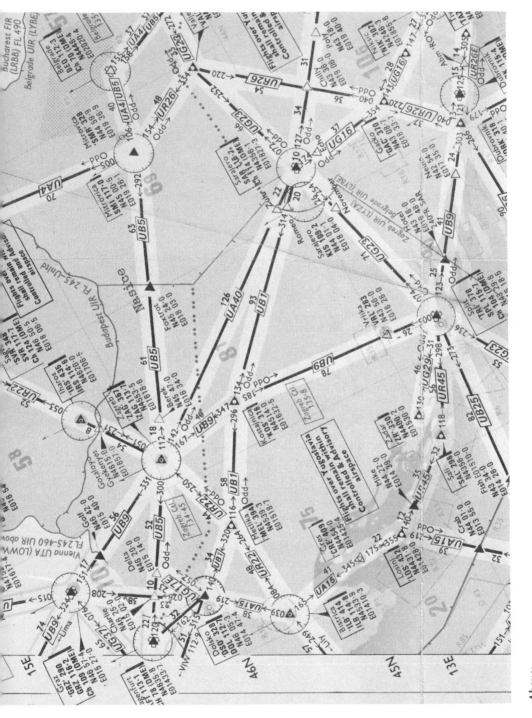

Above:
Detail of Zagreb from Aerad chart.   *Aerad*

132

Flight JP550 on its northwesterly route towards Cologne and Flight BE476 on its southeasterly course towards Istanbul, were both flight-planned to traverse the busy area over the Zagreb radio beacon. A glance at a map of the region clearly indicates Yugoslavia's important position in aviation, as air traffic between Europe and the East routes south of the Eastern Bloc nations. Charter flights from Northern Europe in summer to Yugoslavia, Greece and Turkey also swell the flow of traffic. Five high level airways criss-cross the Zagreb region like a disjointed Union flag with three airways — Upper Blue 5 (UB5), Upper Blue 9 (UB9) and Upper Red 22 (UR22) — intersecting over the Zagreb radio beacon. A fourth airway, Upper Blue 1 (UB1), routes to the south of Zagreb while Upper Amber 40 (UA40), the fifth airway, originates at Zagreb on a direct routeing to Sarajevo. In the previous five years a rapid and sustained growth of air traffic over Western Yugoslavia had resulted in over 760,000 aircraft movements being handled by the Zagreb Air Traffic Control Centre (ATCC). By 1976 the Zagreb Centre was the second busiest in Europe and 30 controllers were struggling to handle traffic which required more than double that number. An extensive training programme had been introduced but the centre was still desperately short of experienced controllers. Three years earlier a modern radar flight control system had been installed but it had not yet been properly calibrated and was considered unreliable. It was claimed that the Swedish equipment did not meet contract requirements. As a result the radar system had not been properly commissioned and was not depended upon for aircraft separation. The Zagreb ATCC relied very much on procedural control with pilots transmitting their positions at specified reporting points along the airways. These position reports were then monitored by the radar system.

The increased work load for the Zagreb staff in the years of air traffic growth produced a very difficult working environment. Lapses in concentration were perhaps not surprising and a number of incidents had resulted, with 32 near-misses being experienced over the last five years. Two controllers had been dismissed for negligence. On a lesser scale lapses in discipline resulted in lateness for duty and unauthorised absence from the control station. These failings, however, had to be viewed in the light of the trying circumstances and in the very high volume of traffic which had passed safely through the region.

On the morning of 10 September 1976, as BE476 and JP550 converged on the Zagreb area, the staff at the Zagreb ATCC faced another day of heavy traffic. The airspace over Zagreb was divided into three distinctly separate layers — lower, middle and upper — with the middle and upper sections each directed in the short-staffed centre by a controller and his assistant, where normally three personnel — radar controller, procedural controller and assistant controller — would be required. JP550, on its climb over the Zagreb radio beacon to its planned flight level of 310, would pass through the middle control layer (from 25,000 to 31,000ft), while BE476, already cruising at level 330, would pass through the upper control section (above 31,000ft). On the duty shift that morning, under the supervision of Julije Dajcic, were five controllers responsible for the middle and upper stratums of airspace. Apart from the 43-year old Dajcic, their ages ranged from late 20s to early 30s. The controllers worked a 12-hour duty day with normally 2hr at a control station followed by a one-hour break. By

10.00hrs GMT (11.00hrs local) the morning shift, which started at 07.00hrs local, had already been on duty for four hours. At the middle section console sat Bojan Erjavec as controller, who had been at his station for the past hour, with assistant controller Gradimir Pelin, who had just started duty. Mladen Hochberger controlled the upper section and was due, at that moment, to be relieved by Nenad Tepes. Hochberger's assistant for the past hour had been Gradimir Tasic who had already spent the first two hours of the shift as duty controller. When Hochberger's relief, Tepes, who was now late, arrived Tasic would also act as his assistant for the next hour, monitoring procedures and co-ordinating flights with other regions on ground telephone links. Of the staff mentioned, all except Tasic, had had a period of at least 24 hours off duty in the last few days. Tasic was on his third day in a row of 12 hours on duty.

As the Inex-Adria flight climbed out of Split, the approach controller who, in spite of his title, also handled airport departures, was having trouble co-ordinating JP550's ascent with Zagreb lower east sector. As a result the Yugoslav DC-9 was required to cross the Split radio beacon at level 120 before continuing its climb to level 190 while proceeding towards the radio beacon at Kostajnica. At 09.55hrs GMT JP550 established contact with the Zagreb lower controller and was further cleared to level 240. One minute later the clearance to climb to level 260, in the middle level airspace, was received by JP550. A request for the DC-9 to call passing level 220 was also made as a reminder for the lower controller to instruct JP550 to change frequency to the middle controller at that point. JP550 had flight planned for level 310 but levels higher than 260 in the middle sector were not available because both levels above were blocked by other traffic. Westbound jet cruising flight levels are 260, 280, 310, 350 and 390 while eastbound levels are 270, 290, 330, 370 and 410. In the middle sector, Adria 584 from Split to Nuremburg was at level 280, estimating Zagreb at 10.08hrs, and Olympic 187 from Athens to Vienna at 310, estimating overhead Zagreb at 10.11hrs. At that time JP550 was estimating Zagreb at about 10.16hrs. At 10.02hrs JP550 radioed passing level 220 and was instructed to transmit the next call on the middle sector's frequency of 135.8MHz. After the frequency change the DC-9 crew waited about 30sec for a break in the flow of conversion then established contact with Erjavec, the middle sector controller.

10.03:21 JP550 R/T: 'Dobar dan [good day], Adria 550, crossing two two five, climbing two six zero.'

10.03:28 Zagreb middle (Erjavec) R/T: '550, good morning, squawk alpha two five zero six, continue climb two six zero.'

10.03:34 JP550 R/T: 'Squawk alpha two five zero six, continue climb two six zero.'

10.03:38 Erjavec R/T: 'That is correct. Inbound Kostajnica, Zagreb, Graz next.'

Alpha 2506 is one of the radar identification codes used in secondary radar operation. With primary radar a signal pulse is simply transmitted from a ground station and reflected back from a target to the station as a weak echo. Secondary radar, on the other hand, involves a ground transmitted radar signal being received on board an aircraft by a small receiver/transmitter known as a transponder, which responds by transmitting a second signal which is in turn received by the ground based radar receiver. The resultant signal on the ground is much stronger than the weak echo detected using primary radar. An added advantage is that the transponder transmits on a frequency different from the ground based radar transmitter. Since the radar receiver is tuned to the transponder frequency, weak echoes of the transmitted radar signal that may be reflected from the target or from storm clouds in the vicinity are eliminated on the radar screen. A distinct clear image of the target is displayed. On instruction from control the transponder is programmed by the pilots selecting a four number code, and the procedure is known as squawking. Each aircraft is allocated its own code which displays on the radar screen by the target a flight label consisting of the squawk code, flight number, and the aircraft flight level. As a further refinement the centre at Zagreb used a radar height filtering procedure where only certain codes were assigned to each control layer. In any airspace layer only aircraft with the appropriate codes (middle sector 2500-2577, upper sector 2300-2377) showed up with a flight label on that controller's radar console. All other aircraft in the vicinity, higher or lower, showed up only as target blips. As a precaution however, if an aircraft strayed unannounced into another layer, for example from the middle to the upper section, then the flight label would in that case appear automatically when the aircraft climbed above level 315. A controller could also obtain the flight label on his screen of an aircraft outside his airspace layer in order positively to identify an unlabelled target. This could be achieved in a number of ways, but the simplest method was to position a 'pointer' on the radar screen over the unidentified blip and then request the computer for information. The identification label would then appear by the selected flight for 30 seconds. The procedure, however, took time, and involved the use of a separate keyboard thus distracting the controller from radar monitoring.

The hand-over of JP550 from the lower east sector to the middle sector during its climb was completed in good time, which was just as well. Contrary to instructions no progress strip with JP550's flight details had been prepared in advance for the middle sector. The flight progress strip is about eight inches by one inch and contains important information on a particular flight including call sign, aircraft type, requested flight level, airway routeing and squawk code. The strips are mounted on metal backing plates which are slotted in racks by the controller. Checking, altering and arranging the strips is part of the assistant controllers' tasks. The Zagreb middle controller had to adjust quickly to accept JP550 into his sector without prior knowledge of the flight, but since the changeover was conducted in the lower airspace the transition was accomplished with safety. The lower controllers conferred well in advance with their middle sector colleagues and the lower east flight progress strip was simply passed on to make up the deficit.

BE476 crossed over the Klagenfurt radio beacon on the Austrian/Yugoslav border at 10.02hrs, about the time of JP550's exchange, and was instructed by Vienna Control to contact Zagreb on frequency 134.45MHz. As the Trident flew eastbound on airway UB5 towards Zagreb at level 330, all was not well at the Zagreb upper control station. Controller Hochberger, still waiting on his relief, was showing his impatience with his colleague's late arrival and vacated his seat to go and look for Tepes. By the time BE476 called Zagreb upper, Tasic, who was supposed to be in the position of assistant, was now completely on his own, controlling flights, monitoring procedures, and liaising with adjacent control centres. Although at 28 Tasic was the youngest in the room he was an experienced and competent controller, but the flow of air traffic through his sector was as much as any man could handle.

10.04:12 BE476 R/T: 'Zagreb, Bealine 476, good morning.'

Zagreb upper (Tasic) R/T: 'Bealine 476, good morning, go ahead.'

10.04:19 BE476 R/T: '476 is Klagenfurt at zero two, three three zero, and estimating Zagreb at one four.'

Tasic R/T: 'Bealine 476, roger, call me passing Zagreb, flight level three three zero. Squawk alpha two three one two.'

10.04:40 BE476 R/T: 'Two three one two is coming.'

Previous radar stations had recorded the Trident level at precisely 330, but on Tasic's screen the flight label indicated BE476 to be at 332 (33,200ft) or 335 (33,500ft). Since the equipment was not considered to be completely reliable the discrepancy was ignored.

10.04:11 TK889 R/T: 'Zagreb, Turkair 889, over charlie, three five zero.'

Tasic R/T: 'Turkair 889 contact Vienna control one three one er . . . sorry, one two nine decimal two. Good day.'

10.04:54 TK889: 'One two nine decimal two. Good day, sir.'

On the Trident flight deck the crew watched the opposite direction Turkish aircraft flash past above them at level 350.

'There he is', called one of the crew.

Tasic now required onward-clearance from Belgrade for an Olympic Airways' flight proceeding eastbound on UB1 towards Sarajevo. He called the Belgrade ATCC on the ground telephone link to speak to the relevant upper controller's assistant.

Tasic Tel: 'I need Sarajevo upper.'

Belgrade Tel: 'Right away?'

10.05:17 OA182 R/T: 'Zagreb, Olympic 182, passing Kostajnica at zero five, three three zero, estimate Sarajevo at one seven.'

Tasic Tel: 'You can hear the message over the phone.'

10.05:20 Tasic R/T: 'Olympic 182, contact. Olympic 182 report passing Sarajevo.'

10.05:25 Belgrade Tel: 'Hello?'

10.05:28 Tasic Tel: 'Hello, hello, listen, give me the controller . . .'

No sooner had the controller taken the phone than another aircraft called.

10.05:30 9KACX R/T: 'Zagreb, Grummen 9KACX with you, flight level four one zero.'

Tasic ignored the call and continued with the telephone conversation.

10.05:35 Tasic Tel: 'Er, Lufthansa 360 and Olympic 182 — they've got nine minutes between them. Is that OK for you?'

Normally aircraft along route should be 10 minutes apart but Tasic hoped the Sarajevo upper controller would accept them with reduced separation.

Belgrade Tel: 'I've got it . . . OK.'

Tasic Tel: 'It's OK?'

Belgrade Tel: 'OK. It's OK.'

The incoming radio calls continued.

10.05:44 IR777 R/T: 'Zagreb, this is Iran Air triple seven. good after . . . morning.'

10.06:15 OM148 R/T: 'Zagreb, Monarch 148, we checked Kostajnica zero five, level three seven zero, Sarajevo one nine.'

10.06:37 9KACX R/T: 'Zagreb, Grummen 9KACX is with you, level four one zero.'

Meanwhile, in the middle sector, JP550 was just levelling at 260. His position was 34nm south of the Kostajnica radio beacon, heading due north on UB9, and estimating overhead Kostajnica at just after 10.09.

10.05:57 JP550 R/T: 'Adria 550, levelling two six zero, standing by for higher.'

10.06:03 Zagreb middle (Erjavec) R/T: '550, sorry three three zero . . . eh . . . three one zero is not available, two eight zero also. Are you able to climb maybe to three five zero?'

10.06:11 JP550 R/T: 'Affirmative, affirmative. With pleasure.'

10.06:13 Erjavec R/T: 'Roger, call you back.'

10.06:14 JP550 R/T: 'Yes, sir.'

On leaving the control room in search for Tepes, Hochberger had met his replacement on the way in and the two men stopped to discuss the air traffic situation. Contrary to instruction they continued to complete the hand-over outside the control room. Tasic, still at the upper control station on his own, was just about managing to handle the volume of traffic although the pressure was beginning to tell. JP550 remained level at 260 while Erjavec attempted to attract Tasic's attention for clearance to climb the Yugoslav DC-9 to level 350 in the upper sector. Climbing aircraft through the layers of airway traffic travelling in both directions requires careful co-ordination by all concerned. Erjavec was satisfied with the situation in the middle sector, but climbing JP550 up through the opposite direction traffic at 330 and checking separation with other aircraft already cruising at 350 was Tasic's responsibility. Erjavec's raised hand was seen by Tasic who, in an obvious gesture, waved him away. All his attention was needed for his own traffic. The middle and upper control stations were only about half a metre apart so Erjavec's assistant, Pelin, who held a radar licence, simply moved across to Tasic's radar screen. Pelin pointed at JP550's unlabelled target and asked Tasic for clearance. The confused exchange was only a brief moment in the sequence of events but proved to have far reaching consequences for those concerned. As far as Tasic understood the situation he was only being indicated an aircraft in the vicinity of Kostajnica, whereas Pelin assumed climb clearance had been received. He moved back to Erjavec with the all clear.

10.07:40 Erjavec R/T: 'Adria 550 recleared flight level three five zero.'

At the upper control station Tasic struggled on alone.

10.07:45 Tasic R/T: 'Beatours 778, squawk alpha two three zero four.'

10.07:50 BE778: 'Alpha two three zero four coming down, 778.'

Again he was on the line to Belgrade for BE778's clearance, but interrupted the call to reply on the radio.

10.08:26 Tasic R/T: '778 radar contact, continue.'

Belgrade was having trouble retrieving Beatours information. Eventually the details were found and BE778's routeing from London to Istanbul confirmed.

'How do you spell Constantinople?' asked Belgrade, using the old name. This was too much for Tasic and he replied in exasperation with the phonetic spelling of that city's international four-letter designator.

'Lima Tango Bravo Alpha.'

There was still no sign of Tepes taking up the controller's position. In the middle sector Erjavec continued to monitor JP550's climb and confirmed by radar that the DC-9 was approaching Kostajnica. Captain Krumpak was instructed to proceed via Zagreb and Graz and to call passing level 290.

10.09:49 JP550 R/T: 'Zagreb, Adria 550 is out of two nine zero.'

10.09:53 Erjavec R/T: 'Roger, call me passing three one zero now.'

As the Yugoslav DC-9 ascended to 350, the BEA Trident 3B, BE476, cruised at flight level 330 with a true airspeed of 480kt. A southwesterly wind aloft at 45kt resulted in a ground speed of 489kt. The estimate for overhead Zagreb remained at 10.14hrs. In spite of the wind direction the Trident had tracked a little to the south of the airway and at 10.11:41 turned left from a heading of 121 degrees on to a heading of 115 degrees to home directly over the Zagreb radio beacon. JP550's rate of climb was 1,800ft/min and the 2,000ft ascent from level 290 to level 310 was achieved in a period of about 1¼ minutes. The DC-9 passed into the upper control sector.

10.12:03 JP550 R/T: 'Zagreb, Adria 550, out of three one zero.'

10.12:06 Erjavec R/T: '550 for further Zagreb one three four decimal four five. Squawk stand-by and good day, sir.'

10.12:12 JP550 R/T: 'Squawk stand-by, one three four four five. Good day.'

Erjavec had instructed JP550 to squawk stand-by on the transponder in order to release the DC-9's middle sector squawk code for another aircraft. Although a not unacceptable procedure, the selection of stand-by laid the foundation of a lethal trap for the unsuspecting Tasic. The radar computer was programmed to activate automatically on the upper controller's screen the flight label of any aircraft transmitting a middle sector transponder code when it climbed above level 315. With the squawk code in stand-by mode no flight label would automatically appear. Tasic's first task on being contacted by JP550 would be to allocate an upper sector transponder code.

On the DC-9 flight deck the frequency was changed, and once again the crew waited for a gap in the flow of radio conversation before attempting to contact Tasic.

10.11:53 Tasic R/T: 'Finnair 1673, go ahead now, copy 1673, go ahead.'

F1673 R/T: 'Finnair 1673 passed Graz at one zero, level three nine zero, estimate . . .'

10.12:10 Tasic R/T: 'Finnair 1673, report passing Delta Oscar Lima. Maintain level three nine zero, squawk alpha two three one zero.'

10.12:20 F1673 R/T: 'Will report passing Dolsko.'

A few seconds before the last Finnair transmission, Tasic received from Pelin JP550's flight progress strip which once again, contrary to regulations, was simply an alteration of the one already held by the middle sector. A new strip should have been prepared by Pelin. Tasic had hardly time to glance at the information. The upper sector station had not received a strip in advance because JP550 was originally flight planned for level 310 which lay within the middle sector. It was not known until the last minute that the DC-9 was sufficiently light to climb to level 350. Unannounced JP550 penetrated the upper sector airspace and showed only on Tasic's radar screen as an unlabelled target.

10.12:24 LH310 R/T: 'Lufthansa 310, Sarajevo at zero nine, three three zero, Kumunovo three one.'

Tasic R/T: 'Lufthansa 310, contact Beograd one three four decimal four five — sorry — sorry, one three three decimal four five. Good day.'

LH310 R/T: 'Good day.'

10.12:38 Tasic R/T: 'Good day.'

Just after Tasic received JP550's flight progress strip Tepes eventually arrived. He had completed the hand-over from Hochberger outside the control room but would need to be updated by Tasic on the latest situation. Tepes was rostered as controller for this duty period so with Tasic at the controller's station they would have to change seats. As the two waited for an appropriate moment Tasic, controlling aircraft in his sector, briefing Tepes in the gaps between and still co-ordinating flights on the telephone with Belgrade, was reaching saturation point. There were eleven aircraft in the upper sector. Having received JP550's flight progress strip at such a late stage Tasic had no time to examine the details or to appreciate the developing situation. A further error confused the situation. Marked on the strip was JP550's level of 350 but omitted was an arrow pointing upwards beside the figures to indicate the aircraft still climbing. As JP550 continued on its ascent to level 350, heading 353° on UB9 towards Zagreb, Tasic was unaware of its height. The DC-9 had still to pass through the opposite direction flight level of 330. The southwesterly wind at that level increased JP550's true airspeed of 440kt to a ground speed of 470kt. At that rate it would cross Zagreb, still climbing, at 10.14hrs, the same time as BE476.

10.12:40 OA172 R/T: 'Zagreb, Olympic 172, good morning, level three three zero.'

10.12:48 Tasic R/T: 'Olympic 172, go ahead.'

OA172 R/T: 'Olympic 172, level three three zero estimate Dolsko at one six.'

10.13:00 Tasic R/T: 'Olympic 172, report passing Dolsko, flight level three three zero. Squawk alpha two three zero three.'

10.13:07 OA172 R/T: 'Olympic 172, squawk two three zero three. Report passing Dolsko level three three zero, and after Dolsko direct Kostajnica?'

10.13:15 Tasic R/T: 'Affirm, sir.'

10.13:18 OA172 R/T: 'Thank you.'

10.13:19 BE932 R/T: 'Zagreb, Beatours 932 is level three seven zero, estimate Dolsko one eight.'

Tasic was distracted again with the attempt to change places with Tepes who was now filling the post of assistant and was in contact with Belgrade on the ground telephone link. Beatours repeated the call and Tasic misheard.

10.13:34 BE932 R/T: 'Zagreb, Beatours 932.'

Tasic R/T: '962, go ahead.'

10.13:42 BE932 R/T: 'Beatours, three seven zero, estimate Dolsko at one eight.'

Once more Tasic's attention was drawn to other matters.

10.13:53 Tasic R/T: 'Beatours, maintain level three seven zero and report overhead Dolsko. Squawk alpha two three three two.'

BE932 R/T: 'Roger, two three three two.'

Still no message had been received from the DC-9 in spite of the short breaks in radio transmissions. The Yugoslav aircraft had 2,500ft left to go to its cruising level of 350 but was only 500ft below the opposite direction Trident at level 330. The two aircraft were both almost overhead the Zagreb radio beacon and were dangerously close. At last the DC-9 made initial contact with the upper controller.

10.14:04 JP550 R/T: 'Dobar dan, Zagreb, Adria 550.'

10.14:07 Tasic R/T: 'Adria 550, Zagreb, dobar dan. Go ahead."

10.14:10 JP550 R/T: 'Three two five crossing, Zagreb at one four.'

Tasic could hardly believe his ears. Did JP550 say crossing level 325? Anxiously he called back.

10.14:14 Tasic R/T: 'What is your present level?'

10.14:17 JP550 R/T: 'Three two seven.'

A possible collision between the Trident and the DC-9 was only seconds away. Neither flight crew was aware of the impending danger but both would be maintaining a normal watch, although spotting another shining aircraft in that bright morning sky would have been very difficult. The Trident was leaving a long contrail which the DC-9 crew may have seen, but judging another's height, especially with the aircraft nose up in a climb is almost impossible. Also the sun was behind the Yugoslav aircraft and eye contact for the Trident crew would have been hard to establish. The ability of the eye to detect other traffic is improved when two aircraft move relative to each other. The important factor of an impending collision is that a line of constant bearing is maintained between two aircraft involved in an encounter and the phenomenon makes it more difficult for them to spot each other. On the Trident flight deck JP550's radio message would have alerted the crew to the danger, but since position reports are normally transmitted after passing a reporting point they may have assumed that the DC-9 was already behind them. Whatever the circumstances, the two aircraft were closing with a speed through the air of around 920kt — at almost 1,100mph: faster than a rifle bullet — and there was almost no time to take avoiding action.

At JP550's last call there was only 300ft and a few seconds separating the flights. It was too late for Tasic to interrogate the computer for the DC-9's flight label as drastic action was required. BE476's flight label incorrectly showed the Trident level at 335. Tasic surmised that if he could hold the DC-9 at its present level of 327 they would pass each other with 800ft to spare. It would be close, but something had to be done. Turning the aircraft on to another heading would take more time and the action of holding JP550 level should have the desired effect. Tasic broke into Croation and, stammering, called back in a desperate attempt to avert a tragedy.

10.14:22 Tasic R/T: '. . . e . . . zadrzite se za sada na toj visini i javite prolazak Zagreba.' (. . . e . . . hold yourself at that height and report passing Zagreb.)

By now the DC-9 was just climbing through 330.

10.14:27 JP550 R/T: 'Kojoj visini?' (What height?)

Captain Krumpak immediately attempted to level the aircraft but for a few seconds its momentum carried it higher.

10.14:29 Tasic R/T: 'No kojo ste sada u penjanju jer . . . e . . . imate avion pred vama na isn . . . ? 335 sa leva na desno.' (The height you are climbing through because . . . you have an aircraft in front of you at . . . ? three three five from left to right.)

140

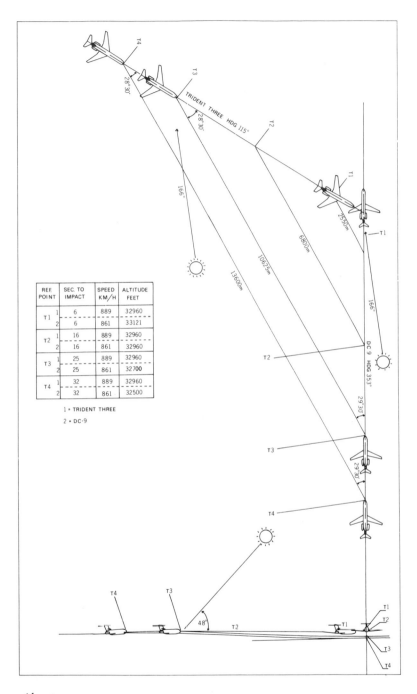

Above:
The final seconds before impact.

141

10.14:38 JP550 R/T: 'OK, ostajemo tocno 330.' (OK, we'll remain precisely at three three zero.)

Unfortunately, in spite of BE476's flight label on Tasic's screen indicating the Trident cruising at 335, the aircraft was flying level at exactly 330. The DC-9 reached maximum altitude of 192ft above the Trident at 10.14:35 then drifted in a gentle descent towards the selected flight level of 330. For a few seconds, as the two aircraft sped headlong towards each other they were both precisely level. Tolerances in radio beacon calibration or altimeter mechanisms, slight variations in headings or wind fluctuations might just have combined to avoid a catastrophe, but it was not to be. At 10.14:41 the outer five metres of the DC-9's left wing sliced through the Trident flight deck as the aircraft collided, killing the crew instantly. The stricken machines dropped out of control from the sky. The DC-9's wing broke off and the left engine disintegrated with ingested debris. As the aircraft descended in flames the empennage detached. The Trident lost its rudder in the collision and dived steeply towards the ground. The cockpit voice recorder on the Yugoslav aircraft had not been working properly but the impact jolted it into action. As the DC-9 tumbled towards the ground it recorded F/O Dusan Ivanus's last words.

'We are finished.'

'Goodbye.' he said, 'goodbye.'

The two aircraft struck the ground about four miles apart near the town of Vrbovec lying about 16 miles northeast of Zagreb. All 54 passengers and nine crew aboard the Trident and 108 passengers and five crew aboard the DC-9 were killed. A total of 176 people lost their lives.

The first civilian airborne collision occurred on 7 April 1922 between a de Havilland 18 (G-EAWO) of Daimlar Airways and a Farman Goliath. The accident happened between Paris and Beauvais, the town associated with the R101, killing the two crew of the UK registered aircraft. History was also made over Zagreb that morning of 10 September 1976, for the mid-air collision between the Trident and the DC-9 proved to be the worst accident of that type on record.

In the few seconds after the crash Tasic, in despair, called both aircraft in turn several times but to no avail. A Lufthansa Boeing 737, travelling eastbound on UB5 at level 290 towards Zagreb, was positioned only 15 miles behind the Trident. The co-pilot 'saw the collision as a flash of lightning and afterwards, out of a ball of smoke, two aircraft falling towards the ground'. The Boeing 737 commander, Captain Joe Kroese, reported the sighting to Erjavec, the middle sector controller.

10.15:40 Capt Kroese R/T: '...e Zagreb! It is possible we have a mid-air collision in sight — we have two aircraft going down, well, almost below our position now.'

Erjavec was unable to understand and Captain Kroese had to repeat his words several times.

10.18:12 Capt Kroese R/T: 'It's possible that the other aircraft ahead of us had a mid-air collision . . . er . . . just overhead Zagreb. We had two aircraft going down with a rapid rate of descent . . . and there was also smoke coming out.'

When the implications of what was being said dawned on Erjavec he glanced across at the upper sector controller. At his station sat a stunned Tasic,

Below:
Distribution of wreckage.

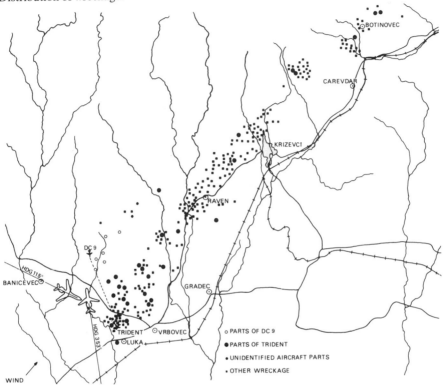

Map labels: BOTINOVEC, CAREVDAR, KRIZEVCI, RAVEN, DC 9, HDG 115°, BANICEVEC, HDG 353°, TRIDENT, LUKA, VRBOVEC, GRADEC

o PARTS OF DC 9
● PARTS OF TRIDENT
▪ UNIDENTIFIED AIRCRAFT PARTS
. OTHER WRECKAGE

WIND

white-faced with shock. Slowly he lifted the head-set from his ears and placed it on the console in front of him.

As the alarm was raised and rescue services were launched, operations ceased in the Zagreb control centre. Incoming aircraft to the airspace were diverted along other airways. The middle and upper sector controllers involved in the collision were suspended, and off-duty staff were summoned from their homes as replacements. Within an hour normal air traffic services were resumed. Soon police and officials arrived at the centre to interview those of the morning shift implicated in the catastrophe: Erjavec, the middle controller who had cleared JP550 to climb; Pelin, his assistant, who had co-ordinated the flight with the upper sector; Hochberger, the upper controller who had vacated his station; Tepes, his replacement, who had been late for duty; Tasic, the upper sector assistant who had been left on his own for up to eight minutes at a critical phase; and Dajcic, the shift supervisor who had held overall responsibility. Within hours they were placed under arrest. At Cologne Airport relatives and friends waited in vain for the arrival of the Inex-Adria charter aircraft. Also kept waiting were the outbound group of tourists checked-in for the return flight to Split. All they were told was that the flight was delayed until 5pm owing to a technical problem. Even 2½ hours after the accident telephone callers were still being given the same facts.

143

It wasn't until early evening that news of the crash was officially released to those waiting, but by then many had already heard of the tragedy from reporters assembling at the airport.

An initial accident inquiry established that the controllers had a case to answer, but the aviation authorities were released from the requirement to publish details. Instead, under Yugoslav law, the controllers also faced charges of criminal misconduct and arrangements were undertaken to place the accused on trial. Within two months all except Tasic were released. He, alone, remained in custody.

Seven months after the collision the hearing of the criminal case opened in the Zagreb District Court before Judge Branko Zmajevic at 8am on the morning of 11 April 1977. The court assembled in the same room where Tito had been tried as a communist in 1923. In the dock sat eight accused: the five controllers and the shift supervisor, plus two senior officials, the head of the Zagreb flight control service, Antere Delic, and the head of Zagreb district flight control, Milan Munjas. Deputy Public Prosecutor Slabodan Tatarac read the long list of charges of criminal negligence.

The trial was well conducted and followed Yugoslav law to the letter. The evidence was carefully presented and formal courtesies observed. The court, however, was not concerned to analyse the cause of the accident, which had been the purpose of the inquiry. It had assembled to apportion blame and to administer punishment. All defendants faced long prison sentences of up to 20 years if found guilty. As the trial progressed the atmosphere at times became heated and tempers flared. Claim and counter-claim followed accusation and counter-accusation. Statements by some of the accused contradicted or embellished evidence presented at the earlier accident inquiry and it became increasingly difficult to establish the facts. Pelin, the middle sector assistant, now claimed that Tasic had been shown JP550's flight progress strip as early as 10.07hrs when co-ordination of that aircraft's climb first took place between the two. Pelin's evidence was supported by Erjavec, the middle controller. None of this had been mentioned in the earlier court of inquiry. In turn Tasic presented fresh revelations. He now recalled that he had told Pelin JP550 could climb to 350 only if it could make level 310 by Kostajnica. If not, it would have to wait for climb until after Zagreb. Pelin denied this exchange. He countered by saying that during the co-ordination of JP550's climb he had also pointed out to Tasic Olympic 182 at Kostajnica level 330, which might have caused conflict with a clearance to climb. According to Pelin, Tasic replied that after they crossed, JP550 was cleared to climb.

The web of information became so tangled that any attempt at establishing the true facts seemed remote. The judge encouraged the accused to challenge each other with the facts but Tasic seldom emerged unscathed from these statements. He became increasingly isolated and fell ill under the strain. Proceedings had to be adjourned for a week.

In court, representing the prosecution, was a British lawyer, Richard Weston. He attended in the capacity of attorney for the parents of one of the stewardesses killed aboard the Trident. His interest, of course, was that justice should be done, but as the proceedings continued he became genuinely concerned with the course

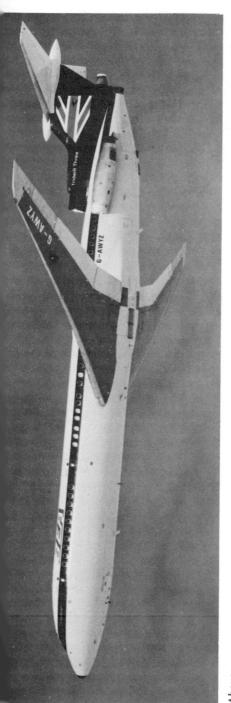

Above:
BEA Trident 3.   *Via John Stroud*

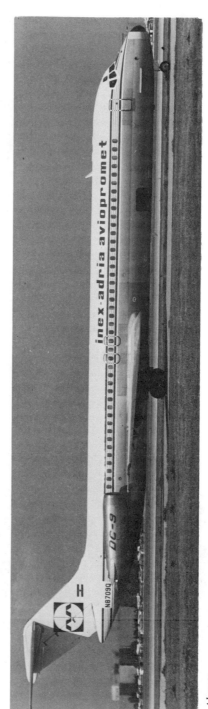

Above:
Inex-Adria DC-9.   *McDonnell Douglas*

Above:
Wreckage with G-AWZ lettering just visible.   *Associated Press*

of the trial. The evidence clearly demonstrated that the air traffic system was as much under scrutiny as the behaviour of the defendants and that changing the system would be far more beneficial than imprisoning the victims of the system. The accused, if found guilty, would be suitably disciplined by the civil aviation authorities anyway, and putting them in jail seemed pointless. Under Yugoslav law, Weston, as an accredited representative, was entitled to be heard by the court. Before the judge's summing-up he decided to exercise that right and, through a Yugoslav colleague reading a translation of his statement, he made an impassioned plea to the court for leniency. Although the controllers may be held responsible, they should not be sentenced to prison like common criminals. It was a most unusual comment from the prosecution benches, but one which was received with applause both within the court room and the world beyond.

On Monday 16 May 1977, the court assembled for the last time to hear the findings read by Judge Zmajevic. In short, of the eight accused, seven were found not guilty. Only Tasic was left to shoulder the blame. The evidence presented by Pelin and Erjavec regarding co-ordination of JP550's climb was preferred to Tasic's. He had made a number of contradictory statements. Tasic had not expressed surprise when given JP550's flight progress strip 1½ minutes before the collision. If he had not given permission to climb why did he not question the hand-over of the strip? On ascertaining the danger of the situation Tasic had neither issued to Captain Krumpak an order with authority nor stipulated an exact height. Had he done so expert witnesses had testified that there was sufficient time for the DC-9 to take avoiding action. The court concluded that Tasic was not overloaded and as such was in a position properly to assess the situation. He was found guilty of criminal negligence and was sentenced to seven years imprisonment.

An appeal was launched immediately and was supported by the sympathetic services of the International Federation of Air Traffic Controllers' Associations (IFATCA) who retained Weston on Tasic's behalf. The Supreme Court reconsidered the case on 29 April 1978, and halved the jail sentence. IFATCA were determined to stop at nothing short of freedom for Tasic and organised a petition which was presented to Marshal Tito. The petition achieved the desired effect and on 29 November 1978, just over two years after the collision, Tasic was eventually released from detention.

Over the years pressure grew for a re-examination of the facts which were first released after the original accident inquiry but were never made available for public scrutiny. In February 1982 the investigation into the collision was re-opened and in September of that year the results were finally published after a painstaking inquiry. Owing to conflicting statements, the report concluded, it was impossible to establish whether co-ordination of JP550's climb had been effected. If co-ordination had been effected it had been conducted improperly by the middle controllers. It was impossible to establish whether Tasic approved JP550's climb. The absence of Tasic's partner at the upper station was against regulations and his departure remained unnoticed by the shift supervisor in spite of him being responsible for all staff. As a result, by any standard, Tasic had become overloaded as he tried to cope alone.

The Tenerife Disaster 1977

On Sunday 27 March 1977, a bomb exploded in the passenger terminal at Las Palmas Airport, damaging the check-in area and injuring a number of people. A separatist organisation, the Movement for the Independence and Autonomy of the Canaries Archipelago, was believed responsible. The Canary Island group is a Spanish possession lying just off the west coast of North Africa, and is rich in tourism, being a haven of fun and sun for travellers from around the world. The capital, Las Palmas, is situated on the east coast of Grand Canary Island, and a short distance to the west lies another island of the group, Tenerife, with the main town of Santa Cruz and its airport, Los Rodeos.

The explosion at Las Palmas was followed by a warning of the possibility of a second bomb, and as a precaution the authorities closed the airport. As a result all flights were diverted to Los Rodeos Airport on Tenerife. The organisation demanding independence from Spain, making their presence felt at the beginning of the tourist season, had achieved the desired result of gaining publicity for their cause. Unknown to the perpetrators, however, their efforts that day were to be completely overshadowed by future events. The Las Palmas blast killed no one but, unintended by the bombers, triggered a sequence of occurrences which, in the end, led to a catastrophe of horrific proportions.

Pan American Flight PA1736, a Boeing 747 — registration N736PA — on charter to Royal Cruise Lines, departed Los Angeles on the evening of 26 March local date (01.29hrs GMT, 27 March) with 275 passengers for New York and Las Palmas. On the stop-over at JFK a further 103 passengers boarded, bringing the total to 378. The crew of 16 was changed. At 07.42hrs (all times GMT) Flight PA1736, call sign Clipper 1736, departed Kennedy for Las Palmas under the command of Captain Victor Grubbs, with First Officer (F/O) Robert Bragg and Flight Engineer (F/E) George Warns. Tending the passengers' needs were 13 flight attendants. About 1¼ hours after the Clipper's take-off from New York, KLM Flight KL4805 departed from Amsterdam at 09.00hrs also en route to Las Palmas. The KLM Boeing 747, registration PH-BUF, was operated by the Dutch airline on behalf of Holland Internation travel group, and on board were 234 tourists bound for a vacation in the Canary Islands, plus one travel guide. After disembarking the passengers, PH-BUF was scheduled as Flight KL4806 to fly back to Amsterdam with an equally large group of returning holidaymakers. The Dutch crew rostered for the round trip was commanded by Captain Jacob van Zanten, a senior KLM training captain, and with him on the flight deck were F/O Klass Meurs and F/E William Schreuder. There were 11 cabin staff to look after the charter passengers.

As both Boeing 747s converged on Las Palmas, the bomb was detonated in the terminal building about one hour before KLM's arrival. With closure of the airport Flight KL4805 simply diverted to Los Rodeos on Tenerife pending the reopening of the airport, and landed at 13.38hrs GMT, the same time as local. At first Captain van Zanten was reluctant to disembark his passengers in case Las Palmas suddenly opened, but after 20min of waiting he reversed his decision and buses arrived to transport the travellers to the terminal. Pan Am's arrival from New York was also affected, but since the opening of Las Palmas seemed imminent and they had adequate fuel, Captain Grubbs asked to hold at altitude near Tenerife. The request was refused and Flight PA1736 also landed at Los Rodeos at 14.15hrs, about 30min behind KLM. The weather was clear and sunny. Many aircraft were diverted from Las Palmas that day and, with Tenerife's own week-end traffic, Los Rodeos was becoming overcrowded. To make use of all available space, aircraft were parked on the runway 12 holding area situated only a short distance from the main apron on the northern side of the runway. KLM was parked nearest the threshold of runway 12 followed in sequence by a Boeing 737, a Boeing 727, a DC-8 and the Pan Am Boeing 747 taking up the rear. These aircraft filled the holding area and subsequent arrivals were parked where space could be found on the main apron. Airport staff worked overtime to handle the doubled amount of traffic.

Meanwhile, the Dutch crew, concerned that there might be insufficient time for them to complete the round trip back to Holland, contacted Amsterdam on high frequency long range radio. A few years before, a Dutch captain was able to extend the crew's duty day at his own discretion according to a number of factors which could be readily assessed on the flight deck. The situation was now infinitely more complex and crews were advised to liaise with Amsterdam in order to establish a limit to their duty day. Captains were bound by this limit and could be prosecuted under the law for exceeding their duty time. Amsterdam replied that if the KLM flight could depart Las Palmas by 19.00hrs at the latest they would not exceed their flight time limitation. This would be confirmed by a message in writing in Las Palmas.

Shortly after Pan Am's arrival, Las Palmas was declared open at about 14.30hrs and aircraft began to depart for Grand Canary Island. Since the Clipper charter

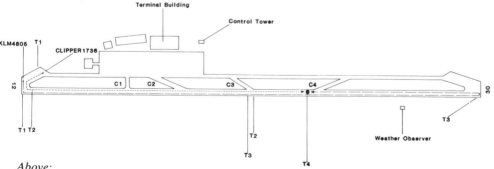

Above:
Tenerife runway details showing relative positions of the aircraft at different times (T1, T2, T3) and point of impact (T4).

flight's passengers were still on board they were prepared for a quick departure. Two company employees joined the flight in Tenerife for the short hop to Las Palmas and were placed in the jump seats on the flight deck, bringing the total on board to 396; 380 passengers and 16 crew. As the aircraft on the holding area positioned between the two 747s were permitted to depart, Pan Am called for clearance in turn. When start-up was requested the controller explained that although there was no air traffic delay they might have to wait. Taxying via the parallel taxi route past the main apron was not possible because of overcrowding, and the holding area exit on to the runway was probably blocked by KLM. The other aircraft parked on the holding area, being smaller in size, had managed to pass behind the KLM 747 on their departure for Las Palmas. Unable to follow suit, N736PA was trapped until the Dutch aircraft moved.

The KLM passengers were summoned from the terminal building, but it took time to bus them back to the aircraft and reboard the flight. Of those who landed in Tenerife only the Dutch company travel guide remained behind, giving a total on board of 248; 234 passengers and 14 crew. In the meantime, many aircraft were arriving at the reopened Las Palmas Airport and it, too, was becoming congested. Los Rodeos was co-ordinating the situation with Las Palmas personnel and soon news came through of a further delay for the Dutch flight. As yet no gate was available for KLM and Captain van Zanten had no choice but to wait. The Dutch crew maintained close contact with Tenerife tower over their departure time and expressed concern over the delay. They had now been on the ground for over two hours. It was unlikely a speedy turn-round could be expected in Las Palmas owing to the congestion, so to ease the situation the Dutch captain made the somewhat late decision to refuel at Tenerife. Since they were still awaiting departure clearance owing to the unavailability of the gate at Las Palmas it would save time on the transit to refuel there for the return Tenerife-Las Palmas-Amsterdam journey. The fuelling process, however, would take about 30min. The Pan Am crew, still hemmed in on both sides, was far from happy with this decision. N736PA was free to leave at any time but could not do so until KLM moved. The American first officer and flight engineer walked out on to the tarmac to measure the distance behind KLM but confirmed that there was insufficient space. Captain Grubbs had been listening in on KLM's radio conversation with the tower and knew of the Dutch captain's desire to leave as early as possible but, unaware of KLM's tight schedule, he felt the Dutch were further delaying both their departures unnecessarily. Clearance for Captain van Zanten to start could come through at any moment but would now have to wait completion of KLM's refuelling. To complicate matters the weather was beginning to deteriorate. Los Rodeos airport is situated at an elevation of 2,000ft in a kind of hollow between mountains and is frequently subject to the presence of low-lying cloud, reducing visibility. In fog, the moisture content and therefore visibility remains relatively constant, but with low cloud drifting across an airport the visibility can change rapidly from several kilometres to zero in a matter of minutes. There was a danger of the weather closing in and preventing departures for a long period. Clouds of varying density from light to dark were blowing down the departure runway 30 with the northwesterly 12 to 15kt wind. At times the runway visibility increased to two or three kilometres and at other times dropped to 300m. The runway centre line lights were inoperative making judgement on take-off more difficult. There was also heavy moisture in the air with the passage of the clouds, and aircraft were frequently required to use their windscreen wipers to clear the view when taxying.

The time was now about 16.30hrs and the Pan Am crew had been on duty for 10¾ hours. They were beginning to feel the strain and were looking forward to their rest after the short 25min hop to Las Palmas. The KLM crew had been on duty for 8¾ hours, but they still had to complete the round trip back to Amsterdam. Three hours remained to the deadline for departure out of Las Palmas, but with the weather worsening this time limit could easily be compromised if they had to wait for the clouds to clear. As it was, the duty limit had not yet been confirmed, and even with fuel on board the transit through Las Palmas could be slow in the overcrowded airport. If the crew ran out of hours and PH-BUF got stuck in Las Palmas, there would be more than a few problems for the ground staff. To begin with, it would be almost impossible to find 250 beds at such short notice and the joining passengers would probably have to spend the night in the airport. The crew would also be late back to Amsterdam and the aircraft would miss the next day's schedules. It was hardly surprising that the Dutch crew were keen to get away. The Americans, too, a little irritated at being held back by KLM, would also be happy to leave Los Rodeos behind.

Permission for KLM to depart did not come through until refuelling was almost finished and vindicated Captain van Zanten's decision to complete the process in Tenerife. As the Clipper crew waited, the American passengers were invited to view the flight deck, and questions about the flight were answered. At 16.45hrs Captain van Zanten signed the fuel log and at 16.51hrs, with pre-start checks completed, KLM requested start-up. Alert to the situation Pan Am heard KLM's radio call.

'Aha', said Captain Grubbs, 'he's ready!'

The Clipper also received start clearance as KLM was starting engines and the two crews prepared to taxi. Owing to the prevailing northwesterly wind, both aircraft had to enter the runway from the holding area for runway 12 and taxi right to the far end of the 11,150ft (3,400m) runway, a distance of over two miles, for take-off in the opposite direction on runway 30.

At that time the control tower had three radio frequencies available of 121.7MHz, 118.7MHz and 119.7MHz. Only two controllers were on duty, however, and 118.7MHz was used for ground taxi instruction and 119.7MHz, the approach frequency, for both take-off and approach communication. KLM was cleared to taxi at 16.56hrs but was instructed to hold short of the runway 12 threshold and to contact approach on 119.7MHz. On establishing contact, Flight KL4805 requested permission to enter and back-track on runway 12. Clearance was received to taxi back down the runway and to exit at the third turn-off, then to continue on the parallel taxiway to the runway 30 threshold. The first officer mis-heard and read back 'first exit', but almost immediately the controller amended the taxi instruction. KLM was directed to back-track the runway all the way, then to complete a 180° turn at the end and face into the take-off direction. The first officer acknowledged the instruction but the Captain, now concentrating on more pressing matters, was beginning to overlook radio calls. The visibility was changing rapidly from good to very poor as they taxied along the runway and it was proving difficult to ascertain position. The approach controller issuing taxi instructions could see nothing from the control tower so was unable to offer any help. One minute later the KLM Captain radioed the approach controller asking if they were to leave the runway at taxiway Charlie 1. Once more KLM were instructed to continue straight down the runway.

Pan Am received taxi instructions on the ground frequency of 118.7MHz and was also directed to hold at the 12 threshold. Captain Grubbs was happy to wait until after KLM's departure but almost immediately they were cleared to follow the Dutch flight. The instruction was given to back-track on the runway and leave by the third exit but because of the ground controller's heavy Spanish accent the crew had a great deal of trouble understanding the clearance. With better communications the Captain might well have established his preference to hold position, but since several attempts were required to comprehend a simple direction it was obviously going to be easier to comply with the controller's wishes. Pan Am was given instructions to back-track on the ground frequency so the Dutch crew, already changed to approach, were at first unaware that the Americans were following behind. As Flight PA1736 began its journey down the runway, ground cleared Captain Grubbs to contact approach. At 17.02hrs KLM now heard the Pan Am crew call approach control and request confirmation of the back-track instruction.

'Affirmative', replied approach, 'taxi into the runway and leave the runway third, third to your left, third'.

By now the Americans had been on duty for over 11¼ hours and were feeling tired. The Dutch crew had been on duty for over 9¼ hours, of which about 3¼ hours had been spent waiting on the ground at Tenerife. The visibility began to drop markedly and fell to as low as 100m in the path of the two Boeing 747s taxying down the runway. It was difficult to spot the exits. Thick cloud patches at other parts on the airport reduced visibility right down to zero. All guidance was given via the radio as no one could see anyone else and Los Rodeos was not equipped with expensive ground radar. Captain van Zanten, when approaching taxiway Charlie 4, again asked his co-pilot if this was their turn off, but the first officer repeated that the instruction had been given to back-track all the way to the end. The KLM crew had switched on the wipers to clear the moisture and suddenly, through the mist, could see some lights. The first officer confirmed the sighting.

'Here comes the end of the runway.'

'A couple of lights to go', replied the Captain. Approach then called asking Flight 4805 to state its position.

17.02:50, Approach R/T: 'KLM 4805, how many taxiway did you pass?'

17.02:56, KLM 4805 R/T: 'I think we just passed Charlie 4 now.'

17.03:01, Approach R/T: 'OK. At the end of the runway make a 180 and report ready for ATC (Air Traffic Control) clearance.'

The KLM crew now asked if the centre line lights were operating and the controller replied he would check. The Americans were still not sure of the correct turn-off because of the language difficulties and once again asked for confirmation that they were to exit the runway at the *third* exit.

17.03:36, Approach R/T: 'Third one sir, one, two, three, third, third one.'

17.03:39, PA1736 R/T: 'Very good, thank you.'
Approach R/T: 'Clipper 1736, report leaving the runway'.

The Clipper replied with his call sign. As the Americans continued down the runway the taxi check was commenced. Instruments and flying controls were checked, the stabiliser was set and the flaps were positioned for take-off, etc. Meanwhile, the Pan Am crew were also trying to spot the turn-offs from the runway in order to count along to the third one, but were having a great deal of trouble in seeing properly. They passed and recognised the 90° taxi exit but were unable to see the taxiway markers so were unsure how many turn-offs they had passed. The allocated exit involved following a 'Z' shaped pattern to manoeuvre on to the parallel taxiway and was going to be difficult for the large Boeing 747 to negotiate. Finding Charlie 3 was also proving to be difficult since its shape was similar to Charlie 2.

As KLM approached the end of the runway the controller called both aircraft regarding the centre line lights.
Approach R/T: 'For your information, the centre line lighting is out of service.'

Each flight acknowledged in turn and checked the minimum visibility required for take-off in such circumstances. By now KLM was commencing the turn at the end of the runway and much was on the Captain's mind. Turning a large aircraft through 180° on a narrow runway requires a degree of concentration and temporarily distracted the captain from other duties. The time was now almost 17.05hrs and the restriction for departure out of Las Palmas was rapidly approaching. If they didn't depart soon they could easily miss it. By good fortune the visibility had improved sufficiently for take-off and with a reduction in moisture the wipers were switched off. If they could get away quickly in the gap in the weather everything should be fine. As the take-off approached, the captain, having to concentrate on so many items, 'seemed a little absent from all that was heard on the flight deck'. He called for the check list.
KLM F/O: 'Cabin warned. Flaps set ten, ten.'
KLM F/E: 'Eight greens.'
KLM F/O: 'Ignition.'
KLM F/E: 'Is coming — all on flight start.'
KLM F/O: 'Body gear.'
KLM F/E: 'Body gear OK?'

The turn was now almost complete with the aircraft lining up on runway 30.
KLM Capt: 'Yes, go ahead.'

The visibility now improved to 900m, but a cloud could be seen ahead moving down the runway. There was just enough time to get away. The Pan Am aircraft was approaching exit Charlie 3 at this stage, half way along the 3,400m runway, but in the bad conditions was unobserved by KLM. Nothing could be seen of the locations of the 747s from the tower. The two aircraft faced each other unseen in the mist.
KLM F/O: 'Wipers on?'
KLM Capt: 'Lights are on.'
KLM F/O: 'No . . . the wipers?'
KLM Capt: 'No I'll wait a bit . . . if I need them I'll ask.'
KLM F/O: 'Body gear disarmed, landing lights on, check list completed.'

At 17.05:28 the captain stopped the aircraft at the end of the runway and immediately opened up the throttles.

KLM F/O: 'Wait a minute, we don't have an ATC clearance.'

The KLM captain, being a senior training pilot, had a lack of recent route practice and was more used to operations in the simulator where he spent a great deal of his time. In the simulator radio work is kept to a minimum on the grounds of expediency in order to concentrate on drills and procedures, and take-offs are often conducted without any formalities. Such an oversight, although alarming, can perhaps be explained by the circumstances. On closing the throttles the captain replied, 'No, I know that, go ahead, ask'. The first officer pressed the button and asked for both the take-off clearance and the air traffic clearances in the same transmission.

KLM F/O R/T: 'KLM4805 is now ready for take-off and we are waiting for our ATC clearance.'

Pan Am arrived at exit Charlie 3 just as approach began to read back KLM's ATC clearance. Having miscounted the turn-offs, they missed their designated taxi route and continued on down the runway unaware of their mistake. They were still about 1,500m from the threshold and out of sight of KLM. It was now over two minutes since the approach controller's last call to Pan Am requesting him to report leaving the runway, and in the KLM crew's desire to depart, the fact that Pan Am was still in front of them, not having cleared the runway, was being overlooked.

17.05:53 Approach R/T: 'KLM4805, you are cleared to the papa beacon, climb to and maintain flight level nine zero. Right turn after take-off, proceed with heading zero four zero until intercepting the three three five radial from Las Palmas VOR [VHF omnidirectional range radio beacon].' Towards the end of this transmission, and before the controller had finished speaking, the KLM captain accepted this as an unequivocal clearance to take-off and said, 'Yes'. He opened up the thrust levers slightly with the aircraft held on the brakes and paused till the engines stabilised.

17.06:09 KLM F/O R/T: 'Ah, roger sir, we're cleared to the papa beacon, flight level nine zero . . .'

As the first officer spoke the captain released the brakes at 17.06:11 and, one second later, said, 'Let's go, check thrust'. The throttles were opened to take-off power and the engines were heard to spin up. The commencement of the take-off in the middle of reading back the clearance caught the first officer off balance and during the moments which followed this, he 'became noticeably hurried and less clear'.

KLM F/O R/T: '. . . right turn out, zero four zero, until intercepting the three two five. We are now at take-off.'

The last sentence was far from distinct. Did he say 'We are now uh, takin' off'? Whatever was said the rapid statement was sufficiently ambiguous to cause concern and both the approach controller and the Pan Am first officer replied simultaneously.

17.06:18 Approach R/T: 'OK . . .'

In the one-second gap in the controllers' transmission, Pan Am called to make their position clear. The two spoke over the top of each other.

17.06:19 **Approach R/T:** '. . . stand-by for take-off, I will call you.'
Pan Am F/O R/T: 'No, uh . . . and we are still taxying down the runway, the Clipper 1736.'

The combined transmissions were heard as a loud three-second squeal on the KLM flight deck causing distortion to the messages. Had the words been clearer the KLM crew might have realised their predicament but, only moments later, a second chance came for them to assess the danger. The controller had received only the Clipper call sign with any clarity but immediately called back in acknowledgement.

17.06:20 Approach R/T: 'Roger, papa alpha 1736, report the runway clear.'

On this one and only radio call the controller, for no apparent reason, used the call sign papa alpha instead of Clipper.

17.06:30 Clipper R/T: 'OK we'll report when we're clear.'
Approach R/T: 'Thank you.'

In spite of these transmissions the KLM Boeing 747 continued to accelerate down the runway. The words were lost to the pilots concentrating on the take-off but they caught the flight engineer's attention. He tentatively inquired of the situation.

KLM F/E: Is he not clear, then?'
KLM Capt: 'What did you say?'
KLM F/E: 'Is he not clear, that Pan American?'
KLM Capt: 'Oh, *yes.*'

The co-pilot also answered simultaneously in the affirmative and the flight engineer did not press the matter. The KLM aircraft continued on its take-off run into the path of Pan Am.

It is difficult for most people to understand how anyone as experienced as the Dutch captain could have made an error of such magnitude. For those used to regular hours and familiar surroundings, with nights asleep in their own time zone and frequent rest periods, it may be impossible to comprehend. But the flying environment, although for the most part routine, can place great strain on an individual. Constant travelling in alien environments, long duty days, flights through the night and irregular rest patterns can all take their toll. The Dutch crew had been on duty for almost 9½ hours and still had to face the problems of the transit in Las Palmas and the return to Amsterdam. Lack of recent route experience for the captain, especially in these trying conditions, did not help. The pressure was on to leave Los Rodeos as early as possible and the weather was not making it any easier. A gap in the drifting cloud had presented itself and the captain had taken the opportunity to depart. Close concentration was required on the take-off as clouds were once again reducing visibility. At such moments the thought process of the brain can reach saturation point and can become overloaded. The 'filtering effect' takes over and all but urgent messages, or only important details of the task in hand, are screened from the mind. Radio communications, which were being conducted by the first officer, were obviously placed in a low priority in the minds of both the pilots once the take-off had been commenced. The controller's use of papa alpha instead of Clipper — the only occasion that day on which this identification was used — reduced the chances of registering the transmission.

On the Clipper flight deck the crew were sufficiently alarmed by the ambiguity of the situation to comment although they were not as yet aware that the KLM had started his take-off run.

Pan Am Capt: 'Let's get the hell out of here.'

Pan Am F/O: 'Yeh, he's anxious, isn't he.'

Pan Am F/E: 'Yeh, after he held us up for an hour and a half . . . now he's in a rush.'

The flight engineer had no sooner finished speaking when the American captain saw KLM's landing lights appear, coming straight at them through the cloud bank.

Pan Am Capt: 'There he is . . . look at him . . . that . . . that son-of-a-bitch is coming.'

Pan Am F/O: 'Get off! Get off! Get off!'

Captain Grubbs threw the aircraft to the left and opened up the throttles in an attempt to run clear. At about the same time the Dutch first officer, still unaware of Pan Am's presence, called 'Vee one', the go or no-go decision speed. Four seconds later the Dutch crew spotted the Pan Am 747 trying to scramble clear.

KLM Capt: 'Oh . . . .'

The Dutch captain pulled back hard on the control column in an early attempt to get airborne. The tail struck the ground in the high nose up angle leaving a 20m long streak of metal on the runway surface. In spite of the endeavours of both crews to take avoiding action, however, the collision was inevitable. The KLM 747 managed to become completely airborne about 1,300m down the runway, near the Charlie 4 turn-off, but almost immediately slammed into the side of Pan Am. The nosewheel of the Dutch aircraft lifted over the top of Pan Am and the KLM number one engine, on the extreme left, just grazed the side of the American aircraft. The fuselage of the KLM flight skidded over the top of the

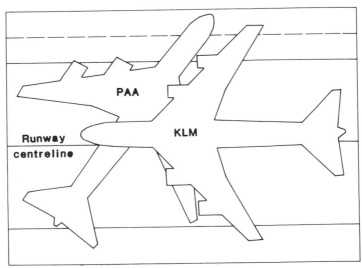

Above:
The position of aircraft at impact.

Above:
A KLM 747 in just about the same altitude that hit the Pan Am 747.   *Via John Stroud*

Below:
Pan Am 747.   *Via John Stroud*

other but its main landing gear smashed into Pan Am about the position of Clipper's number three engine. The collision was not excessively violent and many passengers thought a small bomb had exploded. Pan Am's first class upstairs lounge disappeared on impact, as did most of the top of the fuselage, and the tail section broke off. Openings appeared on the left side of the fuselage and some passengers were able to escape by these routes. The Pan Am aircraft had its nose sticking off the edge of the runway and survivors simply jumped down on to the grass. The first class lounge floor had collapsed, but the flight crew, plus the two employees in the jump seats, managed to leap below into what was left of the first class section and make their escape. On the left side the engines were still turning and there was a fire under the wing with explosions taking place.

The main landing gear of the KLM flight sheared off on impact and the aircraft sank back on to the runway about 150m further on. It skidded for another 300m and as it did so the aircraft slid to the right, rotating clockwise through a 90-degree turn before coming to a halt. Immediately an extensive and violent fire erupted engulfing the wreckage.

The controllers in the tower heard the explosions and at first thought a fuel tank had been blown up by the terrorists, but soon reports of a fire on the airport began to be received. The fire services were alerted and news of the emergency was transmitted to all aircraft. Both 747s on the runway were called in turn without success. The fire trucks had difficulty making their way to the scene of the fire in the misty and congested airport, but eventually the firemen saw the flames through the fog. On closer inspection the KLM aircraft was found completely ablaze. As they tackled the conflagration another fire was seen further down the runway, assumed to be a part of the same aircraft, and the fire trucks were divided. It was then discovered that a second aircraft was involved and, since the KLM flight was already totally irrecoverable, all efforts, for the moment, were

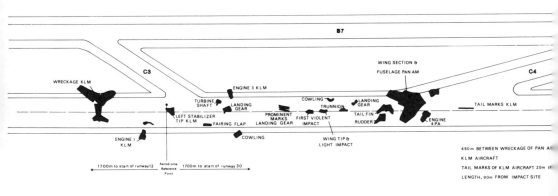

Above:
Distribution of wreckage.

concentrated on the Pan Am machine. Airport staff and individuals who happened to be on the premises bravely ran to help the survivors.

When the extent of the disaster became known a full emergency was declared on the island and ambulances and fire fighting teams were summoned from other towns. Local radio broadcast requests for qualified personnel to offer their services. Although the request for help was made with the best of intentions the rush to the airport soon caused a traffic jam, but fortunately not before the survivors were dispatched to hospital. Large numbers of islanders also kindly donated blood.

Of the 396 passengers and crew aboard the Pan Am flight only 70 escaped from the wreckage and nine died later in hospital. 335 were killed. All 248 aboard the KLM aircraft perished. On that Sunday evening of 27 March 1977, 583 people lost their lives, and as if to mock those in fear of flying, the accident happened on the ground. It stands on record as the world's worst disaster in aviation history.

As the survivors were being tended in Santa Cruz, firemen at the airport continued to fight the infernos on the runway. In the end, in spite of the intense blaze, the firefighters managed to save the left side of the Pan Am aircraft and the wing, from which 15,000-20,000kg of fuel were later recovered. It was not until the afternoon of the following day that both fires were completely extinguished.

Below:
The burnt out tail of the KLM 747 at Los Redeos airport. *Associated Press*

The Chicago DC-10 Crash 1979

American Airlines DC-10-10, registration N110AA, began taxying at 14.59hrs central daylight time (19.59hrs GMT) to the 10,000ft 32 Right runway at Chicago's O'Hare International Airport. The date was 25 May 1979. The weather was clear with fifteen miles visibility and the temperature was 63°F (17°C). A fresh wind blew from 020° at 22kts. The DC-10 was operating Flight 191, a scheduled service from Chicago to Los Angeles, and on board were 258 passengers and 13 crew. On the taxi to the runway the before take-off checks were commenced and the trailing-edge flaps were set to 10°. The leading edge flaps, known as slats, set automatically in sequence. The take-off data card showed V1, the go or no-go decision speed as 139kts, the rotation, or lift-off, speed as 145kt and V2, the safe take-off speed required in the air in the event of an engine failure at V1, as 153kts. All speeds were checked and confirmed correctly set on the airspeed indicator using small bugs to mark the speeds. Approaching the threshold of 32 Right, Flight 191 was cleared into position and instructed to hold. Captain Walter Lux operated the radio while First Office (F/O) James Dillard prepared to fly the sector. The Flight Engineer (F/E) was Alfred Udovich. As the DC-10 moved into position on the runway, take-off clearance was received.

15.02:38 Chicago Tower R/T: 'American 191, cleared for take-off.'
15.02:46 Capt Lux R/T: 'American 191 under way.'

Two seconds later F/O Dillard opened the thrust levers and N110AA, at a gross weight of 379,000lb (just under 172,000kg) began a rolling take-off. F/E Udovich accurately set the power and established take-off thrust by 80kts. During the take-off run left rudder was used to compensate for the strong cross-wind. The machine accelerated rapidly along the runway and at 139kts Captain Lux called 'Vee one.' The DC-10 was now committed to take-off. As the speed continued to increase all systems operated normally. The DC-10 accelerated satisfactorily along the runway until, just two seconds before lift-off, a major emergency occurred. Number one engine seemed suddenly to lose power. At that moment observers on the ground saw smoke and vapour emanating from the vicinity of the engine.

'Damn,' someone shouted in the cockpit.

The pilots were unable to see either the wings or the engines from the flight deck so were unaware of the full extent of the unfolding drama. The left wing-mounted engine and its connecting pylon detached completely from the wing and the whole assembly shot upwards and tumbled backwards over the wing, smashing onto the asphalt. As the engine/pylon unit separated, a three feet section of the wing leading edge was ripped out and vital hydraulic and electrical

160

U.S. crashes

Here is a list of major plane crashes in the United States in the past decade:

■ Nov. 15, 1987, Continental Airlines DC-9 crashed on takeoff from Denver's airport, 28 killed.

■ Aug. 16, 1987, Northwest Airlines MD-80 crashed on takeoff at Detroit Metropolitan Airport, 156 killed.

■ Aug. 31, 1986, Aeromexico DC-9, collided with small plane over Los Angeles suburb of Cerritos, 82 killed.

■ June 18, 1986, De Havilland Twin Otter plane and Bell 206 helicopter, both carrying sightseers, collide over Grand Canyon, 25 killed.

■ Sept. 6, 1985, Midwest Express Airlines DC-9 crashed after takeoff from Milwaukee's Mitchell Field, 31 killed.

■ Aug. 2, 1985, Delta Air Lines Lockheed L-1011 crashed at Dallas-Fort Worth International Airport, 137 killed.

■ Jan. 21, 1985, chartered Galaxy propjet on gambling junket crashed after takeoff in Reno, Nev., 68 killed.

■ July 9, 1982, Pan Am Boeing 727 crashed after takeoff in Kenner, La., 153 killed.

■ Jan. 13, 1982, Air Florida Boeing 737 crashed after takeoff in Washington, D.C., 78 killed.

■ June 2, 1983, Air Canada DC-9 caught fire in flight, landed at Greater Cincinnati International Airport, 23 killed.

■ May 25, 1979, American Airlines DC-10 crashed after takeoff at O'Hare International Airport in Chicago, 275 killed.

■ Sept. 25, 1978, Pacific Southwest Airlines Boeing 727 and private Cessna 172 collided over San Diego, 144 killed.

■ Dec. 13, 1977, a DC-3 chartered from National Jet Service of Indianapolis carrying a college basketball team crashed after takeoff at Evansville, Ind., 29 killed.

Source: The Associated Press

Above:
American Airlines DC-10. *Via John Stroud*

Below:
Firemen battle against the flames. *Associated Press*

lines were severed. As a result important systems were lost, but, owing to a sequence of unfortunate failures, warning devices were inhibited and there was no indication on the flight deck of the additional problems. Unaware of the extent of the damage the crew simply followed standard procedures for engine failure on take-off, which is a well practised manoeuvre in the simulator.

As the engine separated, the lift-off speed of 145kts was reached and the captain called 'rotate'. F/O Dillard pulled back on the control column and at 20.03:38, with 10° nose-up attitude, the DC-10 became airborne. The take-off run had lasted 50sec and used 6,000ft of runway. During rotation the speed had increased through the V2 speed of 153kts and had reached V2+6, 159kts, by lift-off. The DC-10 became airborne with the left wing slightly down and aileron was applied to bring the wings level. The aircraft continued to accelerate and the significance of the left wing dropping, specifically at 159kts, was lost to the crew. Unknown to the pilot, the left outboard leading edge slats had retracted inadvertently, but on the flight deck there was no indication of the event. With such a configuration the left wing tip would stall if the speed were to drop to 159kts and positive control would be required to prevent the aircraft from rolling to the left.

F/O Dillard applied right rudder to compensate for loss of power on the left side and managed to hold the aircraft straight with the heading stabilising at 326°. A number of important flight instruments which would have indicated the situation were supplied by the number one generator electrical busbar and had failed with the loss of number one engine. Power from the other generators should have been relayed to the number one busbar, but a protection system had operated indicating the occurrence of a major electrical fault on the component. In the few seconds after take-off, however, there was little time for trouble-shooting. To reach the electrical and generator reset panel the flight engineer had to reposition his seat, release the safety belt, and get out of the seat.

The first officer's instruments operated normally and he continued to fly the aircraft. American Airlines engine failure procedures called for a V2 climb-out speed of 153kts — 6kts below the left wing tip stall speed — and Dillard pitched the nose-up to the 14° angle displayed on the flight director for a two-engined climb. The vertical speed indicator displayed an initial climb of 1,150ft/min and at 20.03:47, nine seconds after lift-off, the aircraft had accelerated to 172kts at about 140ft above the ground. Aileron and rudder controls maintained the wings level and the heading relatively stable, and the nose-up angle and rate of climb remained constant. With the nose held up at 14° the speed began to decelerate from 172kts to the recommended two-engine V2 climb speed of 153kts. With positive rate of climb established the gear should have been selected up but there was hardly time to take action before the flight began to go seriously wrong. On decelerating towards 153kts the speed fell steadily until at 159kts the left wingtip stalled. There was little or no indication of buffeting, which would have been masked by the light turbulence anyway, and the stick shake system, which would normally alert the crew to a stall situation, had been rendered inoperative. As far as the crew was concerned all the flaps and slats were positioned correctly and the desired V2 climb speed of 153kts was satisfactory. Had they been aware of the impending stall the nose would have been lowered immediately to increase speed

162

TAKE-OFF ENGINE FAILURE
FLAPS 15° OR LESS OR 22°

This procedure assumes indication of engine failure where the take-off is
continued. Each take-off should be planned for the possibility of an engine
failure. Normal take-off procedures ensure the ability to handle an engine
failure successfully at any point.

If an engine failure occurs when making a Standard Thrust take-off, Standard
Thrust on the remaining engines will produce the required take-off performance.
If deemed necessary, the remaining engines may be advance to Maximum Take-Off
Thrust.

Speed . . . . CLIMB OUT AT V_2 UNTIL REACHING 800 FEET AFL
OR OBSTACLE CLEARANCE ALTITUDE, WHICHEVER IS HIGHER,
THEN LOWER NOSE AND ACCELERATE.

At 0°/EXT Min Maneuver Speed,
Flaps . . . . UP

At V_2 + 50 60 ,
Slats . . . . .RETRACT
If returning to land, slats may be left extended.

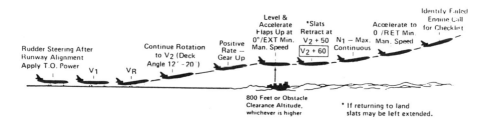

Above:
The AAL engine failure procedure.

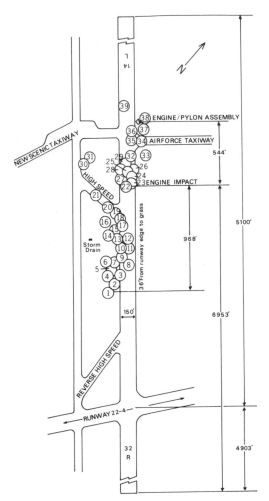

Above:
Distribution of engine/pylon and associated wreckage.
1 Pylon fairing (fibreglass). **2** Seal and torn metal. **3** Pylon fairing (fibreglass). **4** Pylon thrust line shim washer. **5** Flange from pylon thrust link bushing. **6** Pylon pneumatic fitting.
7 Sheet metal AUB 7108-413. Sheet metal ARB 1679-21NC. Steel ferrule. Pneumatic duct clamp. (Piece) 6in dia. ¼in pneumatic line (2½in long). Wing sheet metal ARB 2339-5 with captive nuts. **8** Standoff bracket with two rubber covered clamps. **9** Rib section. **10** Fuel line ARL-0124-23. **11** Control bracket AUN-7020-1. NUN 6001-401. **12** Piece of mono ball (thrust line). **13** Thrust link washer. **14** Wing leading ARB 1065-1G S# Mac 22 (wing L.E. to pylon interface edge with Jay Box). **15** Control bracket AUN 7009-1 (four pulleys).
16 Bolt match to No 21. **17** Sheet metal ARB 1680 (bulkhead type). **18** Fuel line with DBL wall coupling with fuel drain T coupling. **19** Fan cowl ASL 0209-2 (275 707-04). **20** Acoustic material. **21** Sheared bolt and nut ½in dia. **22** Engine pylon impact. **23** Inlet acoustic face sheet piece. **24** Right side cowling pieces, CSD Sv door. **25** Cowling piece. **26** Cowling inlet. **27** Right side fan cowl (piece). **28** Fan blades. **29** Engine drain. **30** Right hand fan cowl. **31** Front spar attach bulkhead piece. **32** Pylon thrust line bushing. **33** CSD generator gearbox. **34** Section of front spar attach bulkhead. **35** Ignition unit. **36** Reverser actuator. **37** Hinged cowling. **38** Control bracket. **39** Fuel line (pylon).

164

and a safe climb-out would have been possible. Instead, the left wingtip began to drop abruptly and the aircraft nose to yaw to the left as the stall developed. With a clearer understanding of the condition it would still have been possible to achieve a safe climb-out, even with an engine failure and the asymmetric slat condition, as long as positive action was taken to maintain the wings precisely level. The minimum speed for controlling the damaged aircraft in the air, with wings level, was as low as 128kts. As soon as the left wing banked 4°, however, the wingtip stall speed of 159kts was the minimum speed at which directional control could be maintained, even with the two operating engines at take-off power.

At 20.03:58, 20 seconds after take-off and with the aircraft only 325ft above the ground, the aircraft began rolling to the left and the left wing banked over 5°. With the speed edging below 159kts, directional control was lost although the situation was not readily apparent to the pilot. Without an increase in speed for which the crew, up to this point, could see no need, the aircraft was beyond recovery. The effect was sudden and catastrophic. The DC-10 continued a rapid roll to the left in spite of large inputs of aileron and rudder in an attempt to maintain directional control. Only 20-odd seconds had passed since lift-off and there was no time to think. The F/O continued to follow the flight director but the nose began to drop, even with full up elevator. At 20.04:06, 28sec after take-off, the DC-10 banked 90° to the left and the nose dropped below the horizon. The wings continued to roll through the vertical and the nose pitched down rapidly to 21°, in spite of full opposite aileron and rudder controls and almost full up elevator being applied. At 20.04:09, after only 31sec of flight and less than one mile from the end of the runway, the left wingtip struck the ground. The aircraft exploded and broke apart, and was completely demolished on impact. Debris scattered over an open field and mobile home park and a fierce fire erupted. Within a short time little remained of Flight 191. All 271 passengers and crew aboard the DC-10 perished and two people on the ground were killed. It was, and still is, the worst accident in American aviation history.

Examination of N110AA's engine pylon recovered from the wreckage on the runway disclosed a 10in fracture on one flange which had originally been caused by over-stressing. Fatigue cracking was also evident at both ends of the fracture indicating the damage had been present for a period of time. During take-off the weakened section had been overloaded and separation of the pylon from the wing had occurred. As a result of these findings the Federal Aviation Administration (FAA) ordered a fleetwide inspection of the DC-10. The checks revealed that a further nine Series 10 aircraft had suffered damage — four American Airlines, four Continental Airlines and one United Airlines — including two Continental Airlines aircraft which had been repaired. Discrepancies in certain clearances and a number of loose, failed or missing fasteners were also found. Examination of the flange fractures established a similarity of the failure modes and prompted an extensive analysis and testing programme of the DC-10 engine/pylon assemblies. The FAA, ever wary of the adverse publicity received following the Paris DC-10 crash, launched a most thorough investigation. The facts which were eventually revealed made disturbing reading.

Evidence of broken fasteners, which did not account for the accident, was found on 31 aircraft. The deficiencies in quality control which had produced these results were traced in part to the disruption caused in 1974 when moving the pylon assembly line from Douglas's factory in Santa Monica to Huntington Beach. Many skilled workers were upset by the change and refused to move. Finding replacements was not easy. In addition to the broken fasteners, the minimal manufacturing clearances at the pylon-to-wing attachment point resulted in maintenance difficulties which made the occurrence of pylon flange damage more likely. This chance of damage was increased by maintenance methods adopted by some airlines.

As early as May 1975, and again in February 1978, McDonnell Douglas issued DC-10 service bulletins requiring maintenance of parts of the engine pylon at the convenience of the operators. The procedure called for dismantling of the complete engine and pylon assembly which was normally accomplished by first removing the engine then detaching the pylon from the wing. American Airlines engineers, while working on four DC-10-30 aircraft in 1977, evaluated the possibility of removing the engine and pylon as a single unit using a modified forklift-type truck. United Airlines was already performing these tasks using an overhead hoist. The technique was calculated to save 200 man hours per aircraft and, from a safety point of view, would reduce the number of fuel, hydraulic and electrical disconnects from 79 to 27. The airline contacted McDonnell Douglas with their proposal, and the manufacturer, with neither the authority to approve nor disapprove their customers' maintenance actions, simply stated that they would not encourage the procedure. At that time, FAA approval of an airline's maintenance methods was not required either, and they were not consulted on the proposed changes. American Airlines, with other DC-10 operators, including Continental Airlines, decided to adopt the process and a considerable number of single-unit engine/pylon assemblies were removed in this manner. The use of a forklift truck to lower and raise the assembly required a great degree of precision on the part of the controller. Post-accident investigations revealed that of the 175 occasions of US airlines removing engines and pylons for maintenance procedures, 88 had been conducted with the engine and pylon as a single unit, 12 of these had been lowered and raised using an overhead hoist and 76 using a folklift truck. All incidents of pylon flange fractures involved use of the forklift vehicle.

In December 1978, Continental Airlines engineers were removing an engine/pylon unit using the forklift truck method when a noise 'like a pistol shot' was heard. An inspection revealed fracture of a pylon flange. The flange was repaired and the aircraft was returned to service. Again, in February 1979, a similar 'pistol shot' was heard when a misunderstanding arose between the forklift operator and the lead mechanic. The nose of the engine was lowered instead of raised, resulting in fracture of the flange. Once more the damage was simply repaired. Continental Airlines examined the incidents and concluded that they were both caused by maintenance errors. The FAA rules governing the reporting of incidents occurring both during maintenance and in service are extensively defined, but in this case the regulations did not seem adequately to outline what constituted a major repair. Also, Continental's mishaps, considered by the airline as maintenance errors, seemed to fall between two categories. As a result there did not seem a requirement to inform the FAA of the incidents and, as in the case

166

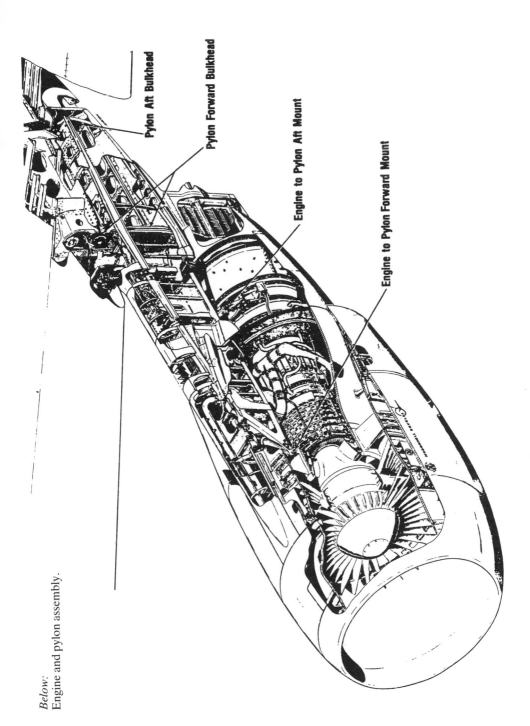

Pylon Aft Bulkhead

Pylon Forward Bulkhead

Engine to Pylon Aft Mount

Engine to Pylon Forward Mount

Below:
Engine and pylon assembly.

of the DC-10 rear cargo door problems, the FAA was totally unaware of the events.

At the occurrence of the Continental Airlines flange fracture in 1978, McDonnell Douglas had sent an engineering specialist to assist in the repair. He was simply informed that the crack had happened when lowering the pylon. The engineer wrote a report on the fracture explaining that it had apparently occurred when the pylon shifted during lowering. An account of this damage was circulated to airlines with a list of other events in a document known as an Operational Occurrence Report, but it was camouflaged by the trivia of other incidents: an air conditioning unit malfunction, a lightning strike, and an injury to a flight attendant. American Airlines did not even recall receiving the Operational Occurrence Report. Once again a major fault had been discovered by one airline but a breakdown in communications had resulted in improper dissemination of the information. American Airlines continued to remove engine/pylon assemblies as a single unit without any knowledge of the inherent dangers.

At the end of March 1979, the accident aircraft underwent modifications. On removal of the left engine/pylon unit it is possible that inadvertent creep down of the lifting truck forks, and/or misalignment of the engine stand on the fork lift, over-stressed the pylon flanges. Unknown to the engineers, a 10in fracture had occurred on one of the flanges. The aircraft had been returned to service in this condition after maintenance and the crack had continued to weaken during the eight weeks of operation before the accident. Finally, the structure had been sufficiently weakened to fail under normal load and the complete engine and pylon had separated from the wing on the take-off out of Chicago.

But why had the aircraft crashed? Engines had failed before, and had even fallen off, as had other aircraft parts, and the crew had been able to recover successfully. What circumstances had caused such tragic results on 25 May 1979? The answer lay in the complications of the DC-10 design philosophy and on the approved engine failure procedures. On separation of the engine and pylon from the wing, three feet of the wing leading edge had been torn off and hydraulic and electrical cables had been severed. The damage had resulted in fracture of the numbers one and three hydraulic lines for the extension and retraction of the left wing outboard leading edge slats. Other aircraft designs included mechanical devices which positively locked out the slats when extended, but the DC-10 did not incorporate such a mechanism. Instead, the DC-10's slats were simply held extended against the air loads by trapped hydraulic fluid in the operating lines. One FAA regulation required symmetric operation of the slats, even with hydraulic problems, and on the DC-10 this was satisfied by a single valve controlling the inboard slats, left and right, and by mechanically linked valves controlling the outboard slats, left and right, which were slaved to the inboard slats. The arrangement ensured uniform operation of the flaps, including taking into account hydraulic failures. In the event, however, it was still considered that asymmetric failure of the outboard slats was a possibility and that a chance remained of the DC-10 being exposed in flight to one set of outboard slats extended and the other retracted. To satisfy the regulations in this respect it was ably demonstrated in flight that with an outboard asymmetric slat condition the DC-10 could be flown safely within the normal range of departure and approach speeds. As an added precaution a slat warning light was introduced which was designed to illuminate when the positions of any slat disagreed with the flap/slat lever position. In addition, a stick shaker was available to operate if early onset of a stall condition, including that induced by asymmetric slat, was detected by the system. If a pilot experienced such an event with the accompanying warning signals, he would simply lower the nose to increase speed and fly out of danger. It seemed an adequate and reasonable design.

According to FAA regulations, any newly designed aircraft was required to be capable of continued flight and landing after 'any combination of failures not shown to be extremely improbable'; the definition of 'extremely improbable' being one chance in a billion. Studies showed that the probability of an uncommanded slat retraction, specifically during the take-off phase, ranged from one chance in a hundred million to two chances in a billion per flight. The consequences, therefore, had to be considered and Douglas complied. An examination of the asymmetric situation at the lower take-off speeds was not conducted in flight although an analysis showed that with all engines operating safe speeds could be achieved. The analysis also showed, however, that with a simultaneous engine failure and asymmetric flap retraction the safety margin might be compromised. A mathematical probability study concluded that the combination was 'extremely improbable' and the design was accepted as complying with the requirements. FAA regulations 'did not require the manufacturer to account for multiple malfunctions resulting from a single failure', and the likelihood of an engine/pylon assembly detaching from a wing, which might have triggered such a sequence, was not considered. McDonnell Douglas deemed 'the structural failure of the pylon and engine to be of the same magnitude as the structural failure of a wing'. Since the loss of a wing, and in Douglas's opinion a pylon assembly, was an unacceptable occurrence, these structures were 'designed to meet and exceed all the foreseeable loads of the life of the aircraft'. The consequences arriving from such failures were not considered. It was logical not to analyse the loss of a wing since continued flight was impossible, but pylon structures had failed in the past and flight had been sustained. An analysis based on the loss of an engine/pylon assembly might have 'indicated additional steps or methods which could have been taken to provide these systems essential to continued flight'. To be fair to the designers, it seems an almost impossible task to predict accurately every likely sequence of events as the result of a failure. In aviation, however, the most unlikely events can and do happen. On 25 May 1979, the seemingly impossible occurred and an engine/pylon assembly detached from N110AA's wing. 'The design and interrelationship of essential systems as they were affected by the structural loss of the pylon' led to catastrophic results.

As the pylon detached hydraulic and electrical lines were ripped out. Normally, when an engine fails, systems continue to function as their power source is automatically switched to the pumps and generators driven by the operating engines. On the DC-10, for example, although the hydraulic systems are independent of each other, the number one hydraulic system, in the event of a number one engine failure, is sustained by its pumps being driven by hydraulic fluid from number two system. Likewise, number one electrical system, in the event of number one engine failure, is sustained by the number one generator busbar being supplied with electrical power from the other engines' generators. The vulnerability of supply lines to a pylon detachment resulted in the tearing out of numbers one and three hydraulic lines. Number one system was lost completely and number three system began discharging fluid, although it continued to operate normally in spite of the leak owing to the short duration of the flight. Had the aircraft continued to fly, only number two hydraulic system would have remained, but aircraft in the past had been landed safely in such a condition. The left outboard leading edge slats were operated by hydraulic lines supplied by numbers one and three systems. Fracture of the slat operating lines had resulted in discharge of the trapped hydraulic pressure locking out the slats, and uncommanded slat retraction had ensued under aerodynamic loads.

The severing of electrical wire bundles within the failed pylon resulted in a short circuit of the number one electrical system. A protective circuitry detected the fault related to the number one generator busbar and automatically isolated the component from other parts of the electrical system. A relay device, designed to supply electrical power from the other generators, tripped and left the number one busbar dead. The flight engineer might have been able to restore power, if the fault had cleared, by operating switches on the electrical and generator reset panel, or by restoring some power by closing the emergency power switch, but the lack of time precluded any such action. With the number one generator busbar isolated, electrical power supply from that source was cut off and a number of items, including some flight and engine instruments, were lost. More importantly, the slat disagree warning light system and the stall warning system were rendered inoperative. The slat disagree warning light remained extinguished in spite of the asymmetric slat condition and resulted in the crew being unaware of the situation. Only one stick shaker system was installed, on the captain's side and, with lack of electrical power to the stick shaker motor, it also failed to function. Even had power been available to the motor the warning would not have activated at the stall speed related to the asymmetric slat condition as the computer sensing the position of the left outboard slat was inoperative as well. As a result no prior warning of the stall condition was received and the only indication the pilot had of the onset of the stall was when the aircraft began an abrupt roll and yaw to the left. The DC-10's two-engine climb-out procedure as practised by a number of airlines and approved by the FAA was intolerant of such an unlikely situation. Crew were trained to fly in such circumstances at a speed equivalent to V2, and the aircraft's flight director was programmed to indicate a nose-up angle at which this speed could be achieved. When DC-10 control was first evaluated, based on the asymmetric leading edge slat situation, it was assumed the appropriate warning devices would be functioning. The separation of the number one engine/pylon assembly, followed simultaneously by uncommanded retraction of the left outboard leading-edge slats and loss of the slat disagreement and stall warning systems, plus an array of failure flags on the flight deck, did not give the crew a chance. As a result of the disaster which befell N110AA, operators were recommended to accept engine-out climb speeds in excess of V2, up to V2+10, so long as obstacle clearance was not compromised.

Three days after the crash the FAA ordered an inspection of engine pylons, and, on 4 June, backed these demands by requiring further inspection of engine and pylon assemblies which had been removed and replaced during maintenance. Results of these inspections revealed damage to engine pylons which greatly alarmed the FAA. On 6 June 1979 the authority's chief administrator, Langhorne Bond, took the unprecedented action of issuing an emergency order of suspension and grounded all DC-10s worldwide. It was feared that the DC-10 design did not meet FAA requirements and the aircraft type certificate was suspended 'until such time as it can be ascertained that the DC-10 aircraft meets the certification criteria'. A total of 270 DC-10s, operated throughout the world by 41 airlines, were affected by the order. The longer range DC-10 (Series 30) aircraft, in which no faults had been found, was also included in the grounding, and McDonnell Douglas and a number of airlines operating the type protested at the FAA actions. A month after the Chicago crash and more than two weeks of grounding, authorities outside the US began to lift the suspension, subject to stringent inspection and maintenance procedures, without American approval. The FAA refused to rescind the order until their own investigation was complete, and countered by issuing an instruction on 26 June prohibiting the 'operation of any Model DC-10 aircraft within the airspace of the United States'. It was not until after five weeks of enforced grounding of American registered DC-10s, on 13 July 1979, that the FAA, satisfied that the DC-10 did meet its requirements and that stricter inspection and maintenance procedures were being implemented, terminated the emergency suspension order and permitted the DC-10s to fly.

The Mount Erebus Crash 1979

In the years 1839 to 1843 a British expedition led by Captain James Clark Ross sailed around Antarctica, exploring the remote regions of that distant and icy continent. In an area which now bears his name, lying almost due south of New Zealand, Ross penetrated the sea of pack-ice to the southern limits of the Pacific Ocean. Sailing down Ross Sea close to the Victoria Land coast, he passed between Ross Island and the mainland to the head of McMurdo Sound, there to be stopped by the Ross Ice Shelf, an impassable frozen barrier about the size of France. On Ross Island, just north of where it meets the permanent Ross Ice Shelf, the explorer discovered two towering volcanoes which he named Terror and Erebus in honour of his gallant ships. Mount Terror, the smaller and inactive volcano, stands at 10,750ft, while Mount Erebus, active and still belching plumes of steam into the icy atmosphere, stands at 12,450ft. Erebus, from the Greek for darkness, signified to those ancient peoples the gloomy region between earth and Hades. It was well named. Almost 140 years after Ross's first exploratory voyages to Antarctica, Erebus was to cast its black shadow over a fine international airline, an entire nation, and much of the world.

In Ross's pioneering footsteps followed other great men like Shackleton, Amundsen, Scott and Byrd, names 'etched' forever in the history of Antarctica. Shackleton first conquered Erebus on 10 March 1908, only a few months before his failed attempt on the South Pole. Three years later Amundsen successfully reached the Pole on 14 December 1911, followed closely by Scott. Near Ross Island Scott's last camp can still be seen preserved for posterity on the ice, where the great trek to the South Pole ended so tragically. In 1929 Byrd took off from a base at the Bay of Whales to become the first man to fly over the South Pole. Today the Ross Sea area and the surrounding land is known as the Ross Dependency, administered by New Zealand. In 1961, however, nations with claims to Antarctic territory signed a treaty declaring the entire region south of latitude 60°S an international preservation area for co-operation in scientific research. Two scientific bases, Scott Base supplied by Australia and New Zealand, and McMurdo Station supplied by the United States, lie about two miles apart on the opposite shores of the long narrow peninsula which joins the southwest corner of Ross Island to the permanent ice shelf. To the southeast of the two bases lies Williams Field, an ice airport operated by the Americans where aircraft of the US Navy, RNZAF and RAAF land supplies for their bases. All air traffic in the area is supervised by US Navy controllers.

Antarctica is a wild, rugged and mountainous land with some of the most spectacular scenery in the world. Ice accumulation on the Antarctic land mass is

6,000ft deep making it the highest continental plateau on earth, with the South Pole standing at 9,000ft. Unlike the flat and featureless frozen sea of the Arctic in the north, the southern ice continent lies remote from the air and sea routes of the world. Apart from a small number of explorers, scientists and military personnel few have ever set eyes upon this region, let alone set foot on her shores. With the advent of cheap and fast jet travel, remote and far off places like Antarctica became a focus of public interest which the airlines were keen to exploit. Air New Zealand first considered Antarctic sightseeing trips as early as 1968 when the proposal to actually land DC-8 aircraft on the ice runway at Williams Field was rejected. Interest was stimulated again in late 1976 when the airline became aware that Qantas was planning flights over the Antarctic from Sydney. Plans to operate two chartered DC-10 services in February 1977 were instigated and the air service licence for the flights duly obtained from the Civil Aviation Division (CAD). The original proposal for flights over the south magnetic pole was soon changed to overflying the more interesting McMurdo Sound area, with the magnificent sights of active Mount Erebus and the Victoria Land glaciers. Flights were planned for the southern summer, with perpetual daylight at the South Pole affording a good chance of a clear view. If low cloud prevailed the south magnetic pole would be an alternative choice.

Antarctica, for all its beauty, is a land fraught with danger, with notoriously fickle weather, even in summer. Flights to the ice would have to be meticulously planned and all contingencies considered. The DC-10 would be unable to land, and the flight from Auckland to overhead McMurdo Sound and back to Christchurch would be an 11-hour ordeal for the crew. Maximum fuel would have to be carried. If any pressurisation problems occurred within the Antarctic region sufficient fuel would be essential for the return journey at very low altitude. Sightseeing in the McMurdo area would have to be at high speed as the risk of lowering flaps for slow flight would be excessive in such cold. If the flaps failed to retract the aircraft would be in an awesome predicament, unable to land on the ice, and unable to return home. Flying low with flaps extended the fuel consumption would be enormous and would rapidly use up reserves. Route structure, minimum safe altitudes within the McMurdo area, sightseeing procedures, communications, whiteout phenomenon and grid navigation practices all required consideration. In polar regions conventional navigation is not possible. The magnetic compass is unusable close to the magnetic pole and reference to the true pole is difficult owing to the rapidly converging meridians. Instead, grid navigation is employed, whereby a grid aligned with the Greenwich Meridian is superimposed on charts and all navigational directions are expressed with reference to the grid. The practice is confusing to the uninitiated with true south becoming grid north thus applying only one north direction over the whole polar area. The principle can be simplified by imagining the mapping graticule found at the intersection of the Equator and the Greenwich Meridian being superimposed over an Antarctic chart. Without such a plan an aircraft proceeding over the Pole would suddenly switch from heading south to heading north (or vice versa) while continuing in the same direction.

The first Air New Zealand DC-10 to overfly Antarctica took off from Auckland on 15 February 1977. The unique opportunity to view the spectacular sights of the

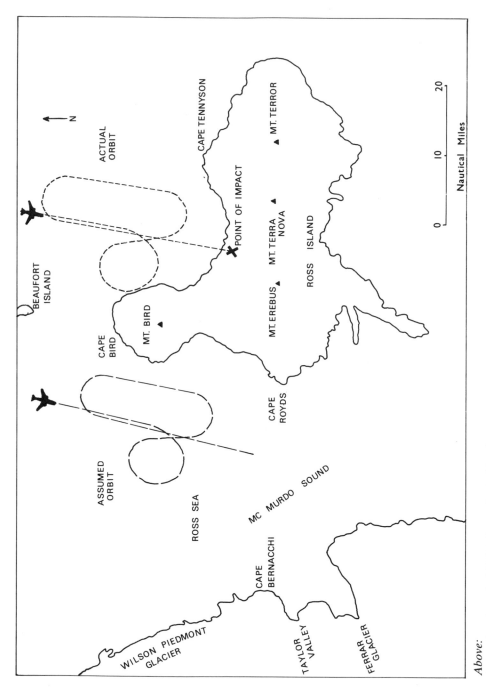

Above:
Antarctica, intended flight plan and actual track.

southern continent was gratefully received by an inquisitive public and from the beginning the trips proved to be popular and successful. The second flight departed later in the month and plans were made for further flights to commence at the beginning of the next southern summer. The commanders of the first two flights were Captain I. Gemmel, Chief Pilot, and Captain P. Grundy, Flight Operations Manager, both of whom at some time during their careers had previously visited Antarctica. This was in keeping with civil regulations and military practice. On the original flight, however, Captain Gemmel's co-pilot was a Captain A. Lawson, at the time the airline's route clearance officer. He was charged with the task of formulating a comprehensive briefing for future crews based on his own experience. The briefing procedure was duly established and, as a consequence, before recommencement of the sightseeing trips in October 1977, dispensation for the requirement of a previous Antarctic visit was requested by the airline and granted by Civil Aviation Division. On all subsequent flights no captain or co-pilot had ever flown to Antarctica before.

On the original flights the final section of the route lay on a direct line from Cape Hallett to a non-directional radio beacon (NDB) close by Williams Field, the ice runway at McMurdo. Overflying of the radio beacon at destination permitted a positive position check of the DC-10's extremely accurate on-board area inertial navigation system (AINS) before commencing sightseeing within the McMurdo area. Such a plan, however, routed the aircraft directly over Erebus requiring a minimum safe altitude for the leg of 16,000ft.

At the commencement of the next southern summer additional flights were undertaken in October and November of 1977. To improve sightseeing on these trips a minimum safe altitude of 6,000ft was established within McMurdo Sound. If cloud conditions existed aircraft were only permitted to descend to this minimum altitude in a specifically defined area to the south of Williams Field, the intention being to break cloud by flying an established let-down pattern from the radio beacon. These instructions remained in force until October 1979 when the NDB was withdrawn. The minimum acceptable cloud base height was 7,000ft for such a procedure, and descent from 16,000ft to 6,000ft was to be monitord by radar. Unfortunately, in the sector designated by the airline, radar coverage above 6,000ft could only be achieved with the greatest of difficulty because of the outmoded equipment in use and was unacceptable to the controllers at McMurdo.

Descent to 6,000ft for cloud penetration at McMurdo 23/10/79

Permission has been given to descend to 6,000ft QNH in VMC conditions, or using the approved NDB procedures in IMC conditions, provided that:

(a) cloud base reported to be 7,000ft or better;
(b) visibility reported to be 20km or better;
(c) ASR is available and used to monitor flight below F/L 160;
(d) no snow showers in the area.

Pilots will have received a comprehensive briefing and completed a simulator detail involving a let-down and climb-out procedure, particular emphasis being placed in the use of grid navigation procedures.

Flight in the McMurdo area below FL 160 will be restricted to an arc corresponding to a bearing of 120°G through 360°G to 270°G from NDB within 20km to keep well clear of the Mount Erebus region.

In practice, however, weather conditions proved good and the visibility excellent in the unpolluted atmosphere. On these and subsequent flights permission was requested and granted from McMurdo ATC for aircraft to depart from the planned track and fly freely in the area. The minimum safe altitude of 6,000ft within the McMurdo Sound sector was based on Mount Aurora standing at just 3,000ft and situated to the south of the sound, well clear of the Williams Field ice runway. Permission was therefore also requested and granted by McMurdo for flights to descend to lower altitudes further to improve the view. Descent below minimum safe altitude is in itself not dangerous as long as aircraft remain clear of high ground. After all, every aircraft in the process of landing, including the military traffic into Williams Field, had to negotiate this process. Permission to descend to altitudes of 1,500ft, the heights flown by military aircraft on the approach to land, was often obtained over the level surface. In addition radar was available, within 40 miles of McMurdo, to monitor flights and to maintain separation with other traffic.

Passengers interviewed after such trips were exhilarated by their experience. New Zealand news media carried reports of the flights and even the Air New Zealand in-house newspaper published a vivid account. The sightseeing flights to Antarctica became a feature of November and trips continued as before except for one notable change in 1978. In August of that year the airline's newly acquired ground computer was being programmed with the company's routes. During this process the destination of the Antarctic flight track was changed from the non-directional radio beacon (NDB) at Williams Field to overhead the Dailey Islands, a small island group situated at the southern end of the sound. This end position was simply named McMurdo. The track change seemed sensible as it avoided overflying the active Erebus volcano. From the beginning that route had appeared less than wise, even though aircraft tended to depart early from track with the McMurdo controllers' permission. But the change of track was unintentional. A small but significant error had been made when one digit of the flight path was entered incorrectly into the computer. The co-ordinates of the NDB (or to be exact, Williams Field) 77° 53.0′ S; 166° 48.0′ E were inserted as 77° 53.0′ S; 164° 48.0′ E. The error had the effect of moving the track about 27 miles west, a fact which went undetected in spite of stringent checks. From this point onwards the computer printed the more logical but incorrect route. This new direct routeing from Cape Hallett to the position over the Dailey Islands, marked as McMurdo, now took Air New Zealand down the centre of the 40-mile wide McMurdo Sound. The flights then passed close to the published military route and the Byrd reporting point over which airforce traffic passed on the landing approach to the ice runway.

On 9 November 1979 a briefing was arranged for the captains flying two of the remaining flights scheduled for November, although other crew members also attended. Of the five present, Captain Simpson, First Officer (F/O) Gabriell and F/O Irvine would take the flight down on 14 November, and Captain Collins and F/O Cassin the flight on 28 November. The briefing was conducted by the route

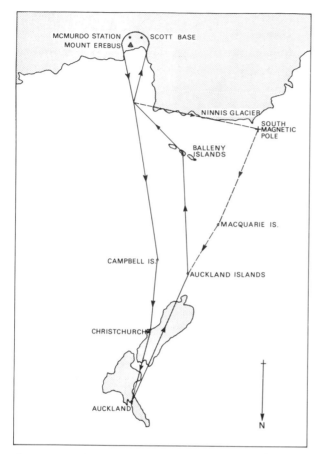

The McMurdo route. The crew's belief that their flight path ran down McMurdo Sound was contributed to by a slide projected at their briefing showing a schematic diagram which displayed a McMurdo Sound route avoiding Mt Erebus. Though an earlier version existed, the slide was produced from artwork prepared by Captain Wilson who briefed the crew on their Antarctic flight.

clearance unit supervisor Captain J. Wilson, a very experienced pilot who had retired into this position. Samples of topographical maps which would be available on the day of the flight were shown but none had the intended route marked. On routine flights crews normally carry charts overlaid with route tracks (usually commercially printed) but such charts were not available for this rather unusual situation. Copies of a number of other maps, however, were available for home study, and each clearly showed the military route passing down the middle of McMurdo Sound to Byrd reporting point. Also available was an actual flight plan from a trip which had been undertaken only two days earlier. On it were typed the latitude and longitude of positions along the route, known as waypoints, which would be inserted into the on-board navigation computer on the day of the flight. Because of the lack of charts indicating the airline route, the exact track and position of the end waypoint at McMurdo were not readily apparent. All these details were examined, however, and notes taken for further study at home. Captain Collins had a small ring-binder notebook, his 'memory' he called it, which he used for such information. A number of slides were also shown depicting views as the flight proceeded southbound. One clearly showed Erebus over to the left of the approach to McMurdo Sound with the taped commentary, 'Now approaching Erebus at 16,000ft, the minimum sector altitude'. Sightseeing procedures, including suitable viewing altitudes, were discussed and authorisation given by Captain Wilson for flights to depart from track and to descend to any altitude approved by the controllers at McMurdo.

McMurdo NDB not available 8/11/79
Delete all reference in briefing dated 23/10/79.
Note that the only let-down procedure available is VMC below FL160 to 6,000ft as follows:

1. Visibility 20km plus.
2. No snow shower in area.
3. Avoid Mount Erebus area by operating in an arc from 120° Grid through 360G to 270G from McMurdo Field, within 20km of TACAN CH29.
4. Descent to be co-ordinated with local radar control as they may have other traffic in the area.

The briefing was conducted in a similar manner to previous occasions except for noting that the radio beacon at Williams Field had been withdrawn. Cloud break procedures using the beacon let-down pattern would not be possible and all descents to lower altitudes would have to be in sufficiently clear skies. The minimum weather conditions of visibility and cloud amount for visual flying are clearly stipulated and are known as visual meteorological conditions (VMC). Circumstances worse than VMC where flying, of necessity, would be on instruments, are known as instrument meteorological conditions (IMC). While flying in IMC, clearly defined instrument flight rules (IFR) apply, resulting in certain specific regulations being mandatory. All aircraft, for example, must adhere to air traffic control (ATC) instructions. Flights at night and in controlled airspace must also comply with IFR, irrespective of the weather. Since most civil transport aircraft operate within controlled airspace they are normally subject to IFR. At certain times when conditions are VMC, however, flights can operate in accordance with VFR, which are also clearly defined but quite different. VFR state that aircraft crews are more or less responsible for their own progress although, in practice, flights still liaise with controllers. The important factors as regards VMC are having sufficient visibility and remaining adequately clear of cloud. As an added precaution Air New Zealand raised the minimum visibility requirements to 20km (12 miles) for Antarctic operations and also stipulated that for VMC descent there were to be no snow showers in the area. Whiteout in falling snow is a well known phenomenon which is especially dangerous at low altitudes and when landing. Sunlight is refracted through ice crystals which turns the sky white, reducing surface definition. People fall over when walking as they literally do not know which way is up.

Past experience of Antarctic flights had shown that VMC normally prevailed and that aircraft, while still in contact with controllers, could sightsee at their will flying freely under VFR in the McMurdo area. Now with the radio beacon inoperative descent would also have to be negotiated in VMC. If VMC was not experienced flights would have to divert to the secondary viewing area to the south magnetic pole. After lunch the group proceeded to the simulator where, under the supervision of Captain R. Johnson, Flight Manager Line Operations DC-10/DC-8, grid navigation techniques were outlined and the VMC descent procedure practised.

On 14 November 1979, Captain Leslie Simpson departed for Antarctica. The weather was fine and afforded good sightseeing for the passengers. The aircraft departed early from track and flew past the glaciers of the Victoria Land coast before turning to fly visually across the sound toward McMurdo Station, descending to 2,000ft. Although the non-directional radio beacon (NDB) at Williams Field was unserviceable, another radio aid used by the Military, the tactical air navigation aid (TACAN), was operating. TACAN transmits both bearing and range data but only range from the transmitter is available to civil aircraft. On crossing the programmed track down McMurdo Sound, Captain Simpson noted from his distance measuring equipment (DME) that the distance to the TACAN was greater than he expected. Suspecting a computer navigation error he overflew the TACAN as a further check but all proved satisfactory. The distance from the TACAN to the McMurdo end of track waypoint was shown to be 27 miles when Captain Simpson assumed it to be nearer 10 miles. On his return Captain Simpson phoned Captain Johnson and mentioned that crews should be apprised of this knowledge in case they too suspected computer error when comparing distances from the TACAN. Captain Johnson, unaware of the incorrect track leading to the McMurdo waypoint over the Dailey Islands, still assumed the track ended at the original waypoint over the now withdrawn NDB at Williams Field. He misinterpreted the remarks and assumed that Captain Simpson was referring to changing the end of track waypoint from the withdrawn NDB to the operating TACAN, a distance of only 2.1 miles. This information was duly passed by Captain Johnson to the navigation section for analysis. The machinery of the airline was slow to respond and the next flight on 21 November left under the command of a Captain White without action being taken. It was not until the following week that the end of track waypoint change was implemented.

On the evening of 27 November, Captain Jim Collins sat down at his dining room table to review the material and notes collected at the Antarctic briefing held nineteen days earlier. He also plotted the co-ordinates of the track waypoints from the flight plan details on a large Antarctic topographical chart which he had purchased. Although this was quite outside normal requirements the conscientious Captain Collins wanted to have no doubt of the route in his own mind. While working, his two elder daughters took an interest and, spreading the chart across the floor, he indicated the track of the next day's flight. After studying the details for about two hours he stopped to watch the 22.00hrs news then retired early to bed. As he slept, work within the navigation section continued through the night.

Tuesday night was the time of the week for updating the stored computer information and Mr L. Lawton, the navigation section superintendent, and Mr B. Hewitt, his chief navigator, were receiving navigational data, including the end co-ordinates of the Antarctic flights. By an unfortunate circumstance the flight plan retrieved for checking was one of the original copies which displayed the direct route from Cape Hallett to the Williams Field NDB. It was therefore a simple task to change the co-ordinates stored in the computer to those pertaining to the TACAN (77° 52.7' S; 166° 58.0' E) and nothing more was thought of it. The change did not seem significant. The effect, however, was to move the end waypoint of McMurdo 27 miles back east to the TACAN near Williams Field, moving the track once again over the top of Erebus. Thus in three years the final waypoint had been changed from longitude 166° 48.0' E (the NDB), routeing seven flights over Erebus in 1977/78, to 164° 48.0' E by mistake, routeing seven flights down the middle of McMurdo Sound close to the military route in 1978/79, then in the early hours of 28 November 1979, by a misunderstanding, to 166° 58.0' E (the TACAN close by the NDB) routeing the planned track back over Erebus. In the course of events the captain of the flight and the controllers at McMurdo would be informed.

On the next morning of 28 November 1979, Captain Collins was up bright and early for the start of the long day which lay ahead. At the airport the flight crew assembled in the dispatch office, checking the weather, fuel requirements, flight logs, ATC clearances and flight notices required for the trip. The normal complement of captain, co-pilot and flight engineer was increased to five, with an extra first officer and flight engineer being required by law for the extra long duty day. With Captain Collins were F/O Greg Cassin and F/O Graham 'Brick' Lucas, and F/E Gordon Brooks and F/E Nick Maloney. A very experienced and competent crew. Of the pilots, however, none had been to Antarctica before and only F/E Brooks had operated on one of the earlier trips. Captain Collins also had extra cabin crew for the flight, amounting to 15 in all, under the charge of Chief Purser Roy McPherson. With the briefing complete, the paperwork was collected for the flight and the crew made their way to the DC-10, ZK-NZP, standing at gate two at Auckland's international terminal. No one at any time mentioned the vital change of the end waypoint co-ordinates or of the significance of the track change. The 'flash ops' message which should have alerted the crew to the change was missing from the flight plan. Somewhere along the line the system had broken down.

Air New Zealand's Antarctic Flight 901 that day was scheduled to be the last of the season and marked the fiftieth anniversary of Commander Richard Byrd's pioneering flight over the South Pole. On board were a total of 237 passengers, one of whom was Peter Mulgrew, a distinguished New Zealand mountaineer and polar explorer. He was there by special invitation and would, from the flight deck, add his own commentary to the sightseeing portion of the trip. While climbing in the Himalayas some years before he had lost both lower legs as a result of frostbite and now walked on artificial limbs. This was not his first trip to Antarctica and he knew the area well. The passengers were mostly New Zealanders with a sprinkling of nationalities from many countries who were on the flight for a variety of different reasons. One New Plymouth policeman had won the tickets in a raffle. As the excited passengers boarded Flight 901 the crew on the flight deck busied themselves with pre-departure checks. Co-ordinates of the route were carefully inserted via a push-button keyboard into the three on-board area inertial navigation systems (AINS), totally self-contained computer navigation units. The AINS is a spin-off from space flights and can be coupled to the autopilot to navigate with extreme accuracy over thousands of

```
                CENTRE LANDING GEAR IS EXTENDED FOR TAKE OFF        ZKNZP ON GATE 2, CLG DOWN FOR DEPARTURE.
OPS FLASH                                                           OPS FLASH
NZN NZAA-NZCH RT NO     /     CAPT DALZIELL     RADIO LOG           NZP NZAA-NZCH RT NO     /     CAPT COLLINS     RADIO LOG
06/11/79-1900Z  TRK.T  W/U  G/S  DIST ZEET FUELRM STN              27/11/79-1900Z  TRK.T  W/U  G/S  DIST ZEET FUELRM STN
M82  TE 901/07  TRK.M DDUUU FL   ZHTA ZETA ROFUEL GMT              M82  TE 901/28  TRK.M DDUUU FL   ZHTA ZETA ROFUEL GMT

NZAA  AUCKLAND                               .  FREQ P             NZAA  AUCKLAND                               .  FREQ P
3700.6S17446.9E                    S/H .... 101.4     S            3700.6S17446.9E                    S/H .... 100.9     S

NP   NEWPLMTH  193.6      400    123   21                          NP   NEWPLMTH  193.6      425    123   20
3900.2S17410.9E 174.3     CLB   .... ....  XX.X                    3900.2S17410.9E 174.3     CLB   .... ....  XX.X

NS   NELSON    199.3 23037 448   146   22    .                     NS   NELSON    199.3 30027 486   146   21    .
4117.8S17308.0E 179.3     FL31  .... ....  91.3                    4117.8S17308.0E 179.3     FL31  .... ....  91.0

RY   MT MARY   216.2 24037 444   208   28    .                     RY   MT MARY   216.2 31027 481   208   26    .
4408.2S17016.8E 195.2     FL31  .... ....  86.5                    4408.2S17016.8E 195.2     FL31  .... ....  86.5

NU   INVRCRGL  211.8 27037 457   163   21    .                     NU   INVRCRGL  211.8 31029 485   163   20    .
4624.8S16819.1E 189.2     FL31  .... ....  83.1                    4624.8S16819.1E 189.2     FL31  .... ....  83.3

AUKIS AKLND IS 198.4 29078 478   271   34    .                     AUKIS AKLND IS 198.4 32029 495   271   33    .
5042.0S16610.0E 173.4     FL29  .... ....  77.5                    5042.0S16610.0E 173.4     FL29  .... ....  77.9

55S   55S      185.7 29098 497   259   32    .                     55S   55S      185.7 31033 498   259   31    .
5500.0S16527.2E 156.2     FL29  .... ....  72.5                    5500.0S16527.2E 156.2     FL29  .... ....  72.9

60S   60S      185.7 31060 504   302   36    .                     60S   60S      185.7 30034 487   302   37    .
6000.0S16431.1E 150.2     FL33  .... ....  66.8                    6000.0S16431.1E 150.2     FL33  .... ....  66.9

BLYIS BALENYIS 185.7 31053 504   407   48    .                     BLYIS BALENYIS 185.7 29026 481   407   51    .
6645.0S16300.0E 349.5     FL31  .... ....  59.6                    6645.0S16300.0E 349.5     FL31  .... ....  59.3

CPHLT C HALLET 155.8 31063 532   367   41    .                     CPHLT C HALLET 155.8 29021 490   367   45    .
7220.0S17013.0E 322.4     FL31  .... ....  53.6                    7220.0S17013.0E 322.4     FL31  .... ....  52.8

MCMDO MCMURDO  188.9 34054 517   337   40    .                     MCMDO MCMURDO  188.5 24015 463   336   43    .
7753.0S16448.0E 357.4     FL35  .... ....  47.9                    7752.7S16658.0E 357.0     FL35  .... ....  46.5

CPHLT C HALLET 008.9 34054 425   337   47    .                     CPHLT C HALLET 008.5 24015 483   336   42    .
7220.0S17013.0E 177.4     FL33  .... ....  41.5                    7220.0S17013.0E 177.0     FL33  .... ....  40.8

70S   70S      358.8 33060 420   139   20    .                     70S   70S      358.8 29024 465   139   18    .
7000.0S17003.6E 168.9     FL33  .... ....  38.8                    7000.0S17003.6E 168.9     FL33  .... ....  38.4

65S   65S      358.8 31068 425   300   42    .                     65S   65S      358.8 29024 465   300   39    .
6500.0S16946.6E 168.7     FL33  .... ....  33.2                    6500.0S16946.6E 168.7     FL33  .... ....  33.3
```

Above:

The flight plan — left, that was produced to Captain Collins at his briefing; right, the altered flight plan delivered on the morning of the flight.

miles without the aid of ground stations. The waypoint data keyed into the AINS was carefully cross-checked for accuracy against the flight plan co-ordinates. Checking the digits of the computer generated flight plan was not part of crew procedures. It was assumed that the computer would produce every time on demand an identical copy of the one and only plan stored in the computer. The alteration of only four digits would not be noticed amongst the mass of other figures. Unknown to anyone on board, the AINS, by an insidious series of errors, was now programmed to fly towards Erebus. The minds of the flight crew, however, were programmed by briefings, maps, previous flight plans, and, certainly in Captain Collins's case, by actual plotting of the track, to fly down the centre of the 40-mile wide McMurdo Sound.

Before departure the flight plan details were transmitted to the military personnel at McMurdo Station. At the beginning of each southern summer they were in the habit of plotting the track of Flight 901 on their own charts. Perhaps, by careful scrutiny, they would spot the change. By a further mistake, however, the flight plan transmitted by the airline to McMurdo had the TACAN co-ordinates omitted and simply displayed the word McMurdo in their place. The controllers at McMurdo Station, assuming the track remained unchanged, were also still programmed to accept Flight 901 down the centre of the Sound. Fortunately the weather forecast for the Antarctic region on the day of the flight was fair with some patchy low cloud but good visibility so it would not cause a problem. Even if, on the way, cloud did materialise to cover the area, the aircraft would simply remain at the minimum safe altitude of 16,000ft. And if visual conditions were good and suitable for descent, what could possibly prevent the crew from seeing the mountain ahead? What indeed.

At 08.17hrs local New Zealand time (19.17hrs GMT Tuesday — local New Zealand time GMT + 13 hours, including one hour summer daylight saving) DC-10 Zulu Papa pushed back from the gate and took-off for the five-hour flight south to McMurdo. The aircraft departed close to its maximum weight of 250 tonnes, with 109 tonnes of fuel for the journey. Climbing smoothly to 35,000ft, Flight 901 proceeded over the Southern Alps of South Island to Invercargill on the coast, then on to overhead the Auckland Islands lying 450 miles from the southern shore of New Zealand. From there the flight turned almost exactly due south for the Balleny Islands, 1,100 miles away. On the journey Captain Collins introduced the crew over the public address (PA) system and explained the sightseeing procedure, asking those by the windows to share seats so all could absorb the view. The latest McMurdo weather reports were relayed from Oceanic Control at Auckland forecasting better weather and the good news was relayed by Captain Collins to the passengers. On board there was a relaxed and party atmosphere, and off-duty crew mingled with the passengers while other travellers visited the flight deck. Everyone received complimentary brochures outlining Antarctic history and after a champagne breakfast a number of documentary films were shown of which one was entitled 'The Big Ice'. In spite of the festive nature of the trip, however, the risks of Antarctic flying could not be overlooked. Only a few weeks earlier the 10th Antarctic Treaty Conference held in Washington noted that such sightseeing flights 'operated in a particularly hazardous environment' and exceeded 'existing capabilities for ATC, communications and search and rescue.' On board Flight 901, for example, there was no polar survival equipment, the airline deeming it unnecessary since an ice landing was not contemplated.

Crossing 60°S the flight crew prepared for grid navigation techniques, aligning the gyro compasses with grid. Tracking almost due true south they would now be heading grid north. By now Flight 901 was in contact with the US Navy controllers at McMurdo air traffic control centre (ATCC), known as Mac Centre, and received reports that weather in the area was not as good as expected. The cloud was down to 3,000ft over Ross Island and the McMurdo area, but with the visibility below cloud an excellent 40 miles. Captain Collins's decision to continue or divert to the south magnetic pole had to be made by the Balleny Islands and there is little doubt that a conference would have ensued amongst the crew and Peter Mulgrew as to their course of action. The captain would certainly be under some pressure to fulfil his obligation to the passengers. After assessing the situation the decision was made to continue.

Suddenly sea ice appeared below shrouding the vista in just two vivid colours of deep blue and brilliant white, to be broken only occasionally by the black of a rocky outcrop or a break in the pack ice. At the Balleny Islands, sitting astride the Antarctic Circle — marked on charts by a broken boundary indicating the extent of perpetual daylight in the southern summer — the navigation computer turned the aircraft towards Cape Hallett. The islands could be seen directly below. The aircraft's present position matched accurately with the sighting of the islands and gave a positive check of the integrity of the AINS.

The DC-10 continued towards the Cape, then rounded the point for the 390 miles leg down to McMurdo. Approaching 200 miles north of the Sound the passengers were informed by Captain Collins that descent would be commenced shortly and the intention was to descend below cloud and enter McMurdo underneath the layer. Before going down Captain Collins conducted a thorough crew briefing, outlining his intentions of following the computer track into the McMurdo area below cloud before commencing any sightseeing activity. If conditions in the Sound were unsuitable it would be wise to climb out of the region at about Byrd reporting point, 23 miles to run to the end waypoint. The aircraft was flying in clear air with the Victoria Land glaciers on the right. Ahead Erebus and the other Ross Island peaks were obscured by cloud at 15,000ft.

At 12.17hrs Ross Dependency Antarctic time (12 hours ahead of GMT — for convenience Antarctic local mean time [LMT] is used throughout) the 'before descent' check was completed. All that remained was for the altimeters to be reset on descent to the local area pressure setting. Approaching 150 miles north of McMurdo the decision was made to commence descent shortly.

12.17:13 Captain Collins: 'I think we'll start down early here.'

F/O Cassin: 'OK, I'll see if I can get hold of them on VHF.'

So far communications had been on high frequency (HF) long range radio, but Flight 901 was now within the 200-mile range of very high frequency (VHF) transmissions. HF radio is subject to static, communications within the frequency band can often be difficult, and messages have to be clear and concise. VHF, on the other hand, is static free and communications are easier. Attempts at VHF contact failed but did not give undue cause for alarm. Perhaps the ground transmitter was operating on low power or the signals were being affected by

Antarctic atmospheric conditions. Mac Centre at McMurdo Station called back on HF with the latest weather.

12.18:05 Mac Centre HF R/T: 'We have a low overcast in the area at about 2,000ft and right now we're having some snow, but visibility is still about 40 miles and if you like I can give you an update on where the cloud areas are around the local area.'

12.18:29 F/O Cassin R/T: 'Yes, 901, that would be handy. We'd like to descend to flight level one six zero.'

12.18:41 Mac Centre HF R/T: 'Kiwi 901, Mac Centre, descend and maintain flight level one eight zero.'

12.18:52 Mac Centre HF R/T: '901, this is the forecaster again. It looks like the clear areas around McMurdo area are at approximately between 75 and 100 miles to the northwest of us but right now over McMurdo we have a pretty extensive low overcast. Over.'

12.19:14 F/O Cassin R/T: 'Roger, New Zealand 901, thanks.'

Ahead they could see the clouds marking the shores of McMurdo Sound.

12.19:22 F/E Maloney: 'That'll be round about Cape Bird, wouldn't it?'

F/O Cassin: 'Right, right.'

12.19:39 F/E Maloney: 'Got a low overcast over McMurdo.'

Captain Collins: 'Doesn't sound very promising, does it?'

The others agreed.

12.19:56 Mac Centre HF R/T: 'Within a range of 40 miles of McMurdo we have radar that will, it you desire, let you down to 1,500ft on radar vectors. Over.'

12.20:07 F/O Cassin R/T: 'Roger, New Zealand 901, that's acceptable.'

Captain Collins: 'That's what we want to hear.'

The three thrust levers of the big jet were eased back and the nose pitched down smartly for descent to flight level 180. Meanwhile Captain Collins lifted the PA handset and spoke again to the passengers of his intentions adding that, although the weather can change quickly 'we're hopeful we'll be able to give you a look at McMurdo today'. This distance on the AINS indicated 114 miles but there was still no range information from the TACAN displayed on the distance measuring equipment (DME) to confirm the accuracy. As the aircraft descended the autopilot was still locked into the AINS which earlier land sightings had shown to be navigating satisfactorily. If anything, it would only be two to three miles in error. The pack ice could still be seen below some scattered cloud as Flight 901 flew on through the clear air above. The areas completely free of cloud could be seen ahead. At this point F/E Maloney relinquished the engineer's position to F/E Brooks, but remained on the flight deck observing. F/O Lucas stayed off the flight deck throughout, viewing the scene from another window. Further attempts at VHF contact with Ice Tower at Williams Field were made without success. The estimated aircraft weight of 119.5 tonnes at 1,500ft was checked and the minimum speed required of 252kt confirmed. A round speed of 260kt indicated would be suitable for flight within the McMurdo area. At 12.24:51 the aircraft levelled off at flight level 180. The crew confirmed AINS navigation track selection and height automatically captured by the autopilot. On HF, Mac Centre called again to suggest re-attempting VHF contact at 80 miles, indicating perhaps that problems were being encountered with their transmitter. As the distance indicated on the AINS, further efforts at VHF contact were to prove in vain. Several times Cassin and Collins tried the two VHF frequencies of 134.1MHz and 126.2MHz on both sets, but to no avail. The problem, however, was not equipment, but terrain. VHF transmissions, TACAN signals, and radar pulses are all 'line of sight'; the radio emissions can pass through most buildings or structures but not through mountains or over the horizon. The AINS programmed track inexorably led Flight 901 towards Erebus, placing the mountain between the transmitters and the aircraft and blocking all signals. HF signals, on the other hand, reflect from ionised layers in the sky and bounce back to earth, so are not affected by terrain.

At this stage it seemed pointless to continue at flight level 180 towards McMurdo Sound. Cloud layers covered Erebus and Ross Island right down to a base of 2,000ft over Williams Field. The segment to the south of the TACAN designated for descent in the McMurdo area was not suitable under the present circumstances, irrespective of the fact that it would not even be considered by the McMurdo controllers because of the radar problems. The sound was mostly free of cloud but the layers covering Ross Island might infringe VMC requirements and might reduce visibility below 20km. There were also snow showers at Williams Field. Although visual descent for Flight 901 required only co-ordination with radar control, and not radar cover, even that was proving difficult. As far as the crew was concerned the only contact with McMurdo was via HF with Mac Centre at McMurdo Station. The radar controllers were situated in Ice Tower at Williams Field, about two miles south of McMurdo Station, with only VHF available. Contact with the radar controllers would have to be via the HF link with Mac Centre personnel who would then have to pass messages to Ice Tower by way of their own link. A most unsatisfactory procedure. There was also no source of bearing signal as an accurate check of the AINS. Equipment to receive TACAN bearings was not available to civil aircraft and the NDB had been notified as having been withdrawn. This information was, in fact, incorrect, and the radio beacon was operating. The Americans had rescinded their original decision and had left the NDB transmitting until it failed. The crew were never apprised of this knowledge and were now unlikely to select a radio beacon which they believed to be unavailable.

At about 60 miles north of McMurdo, as indicated by the AINS, the thin 10,000ft layer below dispersed and large clear areas were now available. Descending in these open spaces well away from the cloud at Ross Island seemed the best course of action, and the aircraft could then enter the Sound below the cloud layer to be picked up by radar at 40 miles. Had the DC-10 been in the process of landing the same procedure would have been followed. In fact, it was not unknown for military aircraft landing at Williams Field to enter McMurdo Sound using INS before being picked up by radar. Military crews, however, unlike their Air New Zealand colleagues, were permitted to descend in IMC conditions under radar control.

12.31:01 Capt Collins: 'I'll have to do an orbit here, I think.' Looking to the left Collins examined the view.

12.31:08 Capt Collins: 'It's clear out here if we get down.'

F/E Brooks: 'It's not clear on the right-hand side here.'

F/O Cassin: 'No.'

12.31:20 Capt Collins: 'If you can get HF contact tell him that we'd like further descent. We have contact with the ground and we could, if necessary, descend doing an orbit.'

12.32:07 F/O Cassin HF R/T: 'We'd like further descent and we could orbit in our present position which is approximately 43 miles north, descending VMC.'

12.32:08 Mac Centre HF R/T: 'Roger, Kiwi New Zealand 901, VMC descent is approved and keep Mac Centre advised of your altitude.'

12.32:10 F/O Cassin HF R/T: 'Roger, New Zealand 901, we're vacating one eight zero. We'll advise level.'

The area was clear all round now as the aircraft nose dipped in descent. The captain disconnected the autopilot from the AINS track and turned the button-sized heading control knob to initiate a right-hand orbit. Flight 901 was now operating under VFR although still co-ordinating with Mac Centre.

12.34:21 Capt Collins PA: 'Captain again, ladies and gentlemen. We're carrying out an orbit and circling our present position and we'll be descending to an altitude below cloud so that we can proceed to McMurdo Sound.'

At the same time Brooks, leaning forward, spotted the Wilson Piedmont Glacier on the right.

12.34:21 F/E Brooks: 'There's Wilson.'

In the cabin, cameras snapped the black breaks in the icy sea and the shores of the distant land.

12.35:15 F/O Cassin: 'Transponder is now responding.'

Halfway round the turn Flight 901 ventured just sufficiently west of Erebus to permit reception of 'line of sight' signals and passed within radar range. An illuminated light on the radar transponder equipment indicated the radar had picked them up at about 40 miles distance, and clearly indicated the accuracy of the AINS. Maloney commented from the back of the flight deck.

12.35:20 F/E Maloney: 'Still no good on that frequency, though?'

F/O Cassin: 'No.'

Another attempt was made at VHF contact which this time proved successful.

12.35:36 F/O Cassin VHF R/T: 'Roger, 901, you are now loud and clear also. We are presently descending through flight level one three zero, VMC, and the intention at the moment is to descend to one zero thousand.'

To the south they could see a broken layer of cloud lying across track, but most of the cloud appeared to be over Ross Island. Flying above the frozen sea the air was mainly clear. As the aircraft completed the right-hand orbit VHF transmissions were once again blocked by Erebus. Six times Ice Tower attempted to re-establish contact.

12.36:32 F/O Cassin: 'We've lost him again.'

The DC-10 levelled off as the autopilot captured 10,000ft. Radar contact was short lived but was sufficient to reassure the crew of their position. No indication was given by Ice Tower of their imminent danger. Perhaps contact had been too fleeting to catch the controllers' attention. Perhaps it was assumed the DC-10 had already commenced sightseeing. After all, Flight 901 was now operating under VFR with the crew responsible for their own progress.

F/O Cassin: 'I'll go back to HF, Jim.'

186

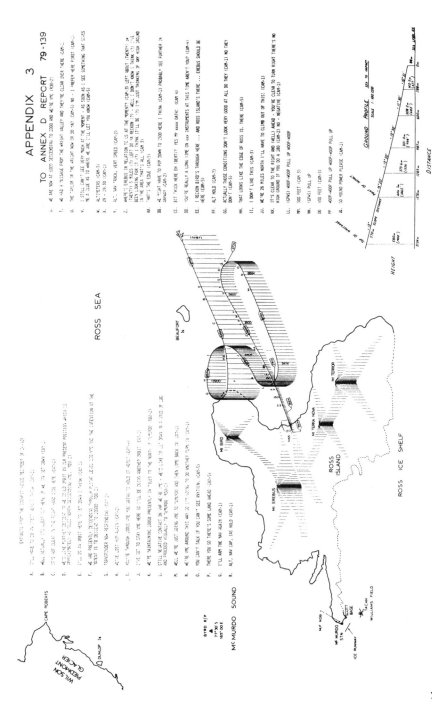

Above:
The last minutes of TE901.

Capt Collins: 'I've got to stay VMC here so I'll be doing another orbit.'

12.38:38 F/O Cassin HF R/T: '901, we briefly had contact on one three four one. We've now lost contact. We're maintaining 10,000ft, presently 34 miles to the north of McMurdo.'

Several more unsuccessful attempts were made on VHF, including a call on the guard frequency of 121.5MHZ. Both sets were checked and the squelch confirmed off.

Capt Collins: 'Tell him we can make a visual descent on a grid of one eight zero and make a visual approach to McMurdo.'

The idea was to descend in VMC in a race track pattern by turning left on to true north for a few miles before turning left again back on track towards McMurdo Sound.

12.42:01 F/O Cassin HF R/T: '901, still negative contact on VHF. We are VMC and we'd like to let down on a grid of one eight zero and proceed visually to McMurdo.'

12.42:05 Mac Centre HF R/T: 'New Zealand 901, maintain VMC. Keep us advised of your altitude as you approach McMurdo.'

Mac Centre also requested Flight 901 to report ten miles from McMurdo. The big jet banked left in a descending turn and headed due north, continuing down to 2,000ft. The crew confirmed the autopilot set to capture 2,000ft altitude and the speed hold engaged. On the right could be seen what appeared to be the western escarpment of Mount Bird. As the DC-10 dropped lower they could see below the edges of the distant cloud layers and visibility was at least 40 miles.

12.42:49 Capt Collins: 'We're VMC around this way so I'm going to do another turn in.'

Collins banked the aircraft once more in a descending left turn to head south and pick up the track once more towards McMurdo. At this point Peter Mulgrew entered the flight deck and manoeuvred into his seat.

12.42:59 Capt Collins: 'Sorry, haven't got time to talk, but . . .'

Mulgrew: 'Ah well, you can't talk if you can't see anything.'

Captain Collins now commented that at 50 miles they were still not picking up VHF. With no line-of-sight signal reception on VHF, TACAN or radar, perhaps alarm bells should have been ringing. The earlier transponder light illumination, however, may have given the impression that radar contact was established and they may also have assumed that at low altitudes they were still out of VHF range. There was also a feeling that the indicated TACAN channel frequency was incorrect and it was re-checked several times. While the aircraft banked to the left large black patches marking breaks in the ice could be seen all around. Perhaps that break in the distance was an island. Or was it just another black patch? It was difficult to tell. If it was land it would have to be Dunlop Island by Victoria Land coast. As they were looking to their left in the direction of turn and were not viewing straight ahead they would have only caught a glimpse. Had Peter Mulgrew taken his seat earlier he may have recognised something, but the area now lay to their right out of view from the flight deck.

12.43:27 Mulgrew: 'There you go. There's some land ahead.'

It looked like Cape Bernacchi out on the right.

Capt Collins: 'I'll arm the nav again.'

The AINS was now re-engaged to navigate the aircraft via the autopilot. The visibility all around now was 40 miles and they could see ahead the ice covered sea stretching into the horizon. Mac Centre came up again on HF and requested reports on the cloud layers.

12.44:47 F/O Cassin HF R/T: 'Roger, New Zealand 901, 50 miles north the base was one zero thousand. We are now at 6,000 descending to 2,000 and we're VMC.'

Jim Collins turned to speak to Peter Mulgrew.

12.45:26 Capt Collins: 'We had a message from the Wright Valley and they are clear over there.'

Mulgrew: 'Oh good.'

12.45:31 Capt Collins: 'So if you can get us out over that way . . . ?'

Mulgrew: 'No trouble.'

12.45:36 Mulgrew: 'Taylor on the right now.'

In the distance he could see what appeared to be the edge of Taylor Valley lying just to the south of Cape Bernacchi. It would not yet be in view for the passengers.

Mulgrew PA: 'This is Peter Mulgrew speaking again folks. I still can't see very much at the moment. Keep you informed soon as I see something that gives me a clue as to where we are. We're going down in altitude now and it won't be long before we get quite a good view.'

As they passed through 3,000ft there was much rustling of maps on the flight deck with the crew checking from charts the edges of land they could see on either side. The sighting confirmed the aircraft approaching down the middle of McMurdo Sound. Had a quick check been made of the latitude and longitude readout displayed continuously above the pilots' heads their true position would have been indicated, but at that time Air New Zealand procedures only required use of the overhead co-ordinates at pre and post flight, or if one system was in disagreement. There was also no reason to doubt the planned track. The figures, however, would have had to be plotted on a chart and the task would have detracted from maintaining visual reference.

12.46:39 F/E Brooks: 'Where's Erebus in relation to us at the moment.'

Mulgrew: 'Left, about 20 or 25 miles.'

F/O Cassin: 'Yep, yep.'

12.46:48 F/E Brooks: 'I'm just thinking of any high ground in the area, that's all.'

Mulgrew checked the position from a map.

Mulgrew: 'I think it'll be left.'

F/E Maloney: 'Yes, I reckon about here.'

Mulgrew: 'Yes . . . no, no, I don't really know.'

Returning to the view he pointed over to the left to cliffs on the coast and confirmed his thoughts.

12.47:02 Mulgrew: 'That's the edge.'

The DC-10 now levelled at 2,000ft with the speed hold engaged and the AINS still navigating on track. The cloud ahead continued to lower to its 2,000ft base.

12.47:43 Capt Collins: 'We might have to pop down to 1,500ft.'

F/O Cassin: 'Yes, OK. Probably see further anyway.'

The aircraft dipped its nose in gentle descent.

12.47:49 F/O Cassin: 'It's not too bad.'

Mulgrew looked about him and checked the chart details against the coast lines which could be seen on both sides. He pointed back to the left.

Mulgrew: 'I reckon Bird's through here and Ross Island there. Erebus should be here.'

At 1,500ft the autopilot captured the height and the DC-10 sped on at 260kts over the sea ice. Further discussion ensued on the problem with the TACAN. All looked ahead for a glimpse of the buildings at McMurdo Station. The sea ice of what appeared to be McMurdo Sound stretched over what could only be the Ross Ice shelf to a distant horizon. As the headlands on either side receded behind the aircraft, however, the horizon ahead began to melt from view.

12.48:46 Capt Collins: 'Actually, these conditions don't look very good at all, do they?'

Mulgrew: 'No they don't.'

There was now concern at the lack of VHF contact and Collins requested Cassin to try again.

12.49:08 Mulgrew: 'That looks like the edge of Ross Island there.'

But it wasn't Ross Island. They were, in fact, flying south into Lewis Bay, the shores of which on either side bore a remarkable resemblance to those expected at the head of McMurdo Sound. Lying in the DC-10's path, obscured from sight by a freak of nature, was the giant Mount Erebus. In spite of the endeavours of the crew to navigate with safety and accuracy, the aircraft was most certainly not where the crew, or the controllers, expected it to be. F/E Brooks, positioned further back than the two pilots, was not happy with the view from his seat.

12.49:24 F/E Brooks: 'I don't like this.'

Visibility, however, was still 40 miles and the passengers were happily snapping shots with their cameras. Sunlight streamed into the cabin.

12.49:25 Capt Collins: 'Have you got anything from him?'

F/O Cassin: 'No.'

Still no VHF reception. It was now several minutes since the last radio contact.

12.49:30 Capt Collins: 'We're 26 miles north. We'll have to climb out of this.'

The intention was to turn and back-track the route. Cassin assured his captain that it was clear on the right if he wished to turn that way. Collins, with Mulgrew, was looking to the left and he preferred to turn in the direction he was looking.

12.49:35 Mulgrew: 'You can see Ross Island? Fine.'

12.49:38 F/O Cassin: 'You're clear to turn right. There's no high ground if you do a one eighty.'

Collins was happier turning left.

Capt Collins: 'No . . . negative.'

The captain pulled the heading select knob to disengage the AINS tracking. The DC-10 adjusted slightly as it settled on the heading already selected on the heading cursor. Collins prepared to turn left. Suddenly the ground proximity warning system (GPWS) sounded a monotone warning. 'Whoop, whoop. Pull up. Whoop, whoop' the electronic voice intoned. The warning must surely be false. They were in the middle of McMurdo Sound with nothing visible ahead. Regulations, however, demanded that such devices be obeyed, and the crew immediately commenced the drill calmly and efficiently. Captain Collins

190

disconnected the autopilot and pulled back on the control column with his left hand, lifting the nose about 15°. Simultaneously he pushed forward the throttles with his right hand. The flight engineer called out the heights on the radio altimeter.

12.49:48 F/E Brooks: 'Five hundred feet.'

GPWS: 'Pull up.'

F/E Brooks: 'Four hundred feet.'

GPWS: 'Whoop, whoop. Pull up. Whoop, whoop. Pull up.'

Collins, in a firm but calm voice, requested the flight engineer accurately to set the engine power.

Capt Collins: 'Go-around power, please.'

GPWS: 'Whoop, whoop. Pull –.'

The DC-10 struck the lower slopes of Mount Erebus and shattered into a million pieces. The underbelly of the aircraft smashed an indentation 4m deep into the ice and snow as the two wing engines chewed into the mountainside. In a small fraction of a second all 257 people on board perished. Instantly a great fire ball erupted. The tail engine ran on a few seconds longer, catapulting the tail fin and sections of the cabin up the slope. The inferno raged for some time, melting the ice beneath and blackening the wreckage. As the last of the flames subsided a dark scar 600m long slashed the snowy whiteness.

Flight 901 had been unseen since the aircraft departed from Auckland in the morning and by late afternoon concern was growing at the lack of contact. Regular news bulletins kept the nation informed of the drama. New Zealand had never before experienced a major air crash and the country began to fear the worst. At 20.00hrs New Zealand local time it was announced on the news that the aircraft's fuel would by now have run out. The DC-10 had to be presumed lost. Perhaps it had ditched in the icy seas? In the perpetual daylight of Antarctica search for the missing aircraft continued into the late hours until suddenly, at 01.00hrs, New Zealand local time, a US Navy Hercules spotted the dark line of wreckage. The morning papers carried the sad story and the news rocked the nation. New Zealand has a small population with close-knit communities, and almost everyone seemed to know someone on the flight. It was the fourth worst disaster in aviation history at the same time and yet another involving the ill-fated DC-10.

Immediately plans were laid to recover the bodies and search the crash site. On the evening of 29 November a Hercules aircraft departed from Christchurch for Antarctica with a number of personnel from the police, Air New Zealand, the press, and the accident investigation branch, including Ron Chippendale, the Chief Inspector of Accidents. American investigators and aircraft company representatives also left the States for McMurdo. Within a few days recovery work was underway and soon the flight data recorder and the cockpit voice recorder tapes were found. By mid-December the results had been analysed and any aircraft malfunction ruled out. McDonnell Douglas personnel, heaving a sigh of relief, hastily released a statement declaring the DC-10 free from blame.

Soon talk of low flying in cloud and mumblings of pilot error surfaced within Air New Zealand and in newspaper articles. There was also a rumour of a flight plan computer error. A directive from the Chief Executive, Morrie Davis, forbade employees to talk to the press and a file was opened by the company for all documents relevant to the crash. All surplus material and flight papers not relating to the accident were to be destroyed. In a number of cases the decision to discard or file data was in the hands of the very airline employees who might be implicated in the disaster. Many papers simply disappeared. Airline personnel called at the homes of the pilots requesting all material relating to the flight. Captain Collins had taken his charts and ring-binder notebook with him but F/O Cassin's papers were collected and returned later with some pages missing, never to be seen again. On 19 February 1980, Air New Zealand issued a public denial that incorrect navigation data had been used for the flight. The statement was accurate in that the originally intended flight plan track had been used, but was otherwise highly misleading. On board were at least 200 cameras and many were recovered undamaged from the snow. A large number of photographs were printed which clearly showed land on both sides of the aircraft. So much for the 'flying in cloud' theory.

By the beginning of March, Chief Inspector Ron Chippendale's interim accident reported was delivered to the Minister of Civil Aviation. Shortly afterwards it was announced that a Commission of Inquiry would be established to investigate the cause of the accident, but would not be convened until after Chippendale's final report. Since the interim report clearly implicated the crew, there seemed little doubt that the pilots would be found at fault. But how was this possible? Jim Collins was well known in the airline for being a conscientious and able captain. How could any pilot in his right mind, never mind one with Collins's reputation, fly at 1,500ft towards a cloud-covered Mount Erebus knowing that his track at that height would take him, not over the top, but into the side of the mountain. It just did not make sense. One pilot above all others, Captain Gordon Vette, placed his career and his future at stake to fight for the good names of his colleagues. He was a good friend of Jim Collins and a close friend of Gordon Brooks and knew both men well. Much of Vette's convincing evidence on visual perception and associated problems was subsequently presented at the Commission of Inquiry and helped lead to a full explanation of the disaster. Later his work was to lead to a book, *Impact Erebus*, published in 1983, and a video film of the same name released in 1985. Each is dedicated to flight safety, with income from both sources being used to fund a flight safety research trust set up by Vette himself.

In Auckland, meanwhile, battle lines were being drawn between two distinct groups of antagonists: on the one side senior Air New Zealand management and executive pilots, prepared to sacrifice a crew to the accusation of pilot error to save the name of the airline, and on the other side colleagues of the deceased flight crew and the New Zealand Airline Pilots' Association (NZALPA). It was to be a bitter confrontation which was to lead all the way to the Privy Council in Britain.

By 20 June, midwinter in New Zealand, the Chief Inspector's final report was ready and would be used as evidence in the forthcoming inquiry set to open in July. There seemed little point in releasing the details at that time but, amid much protest, the report was published. The government, with a vested interest in Air New Zealand, gave top priority to the printing of the report. The result was a

192

foregone conclusion. 'The probable cause of the accident was the decision of the captain to continue flight at low level towards an area of poor surface horizon definition when the crew was not certain of their position, and the subsequent inability to detect the rising terrain which intercepted the aircraft's flight path.' The conclusion was based on five main points:

(1) The change in the computer flight path from McMurdo Sound to Lewis Bay did not mislead the crew.
(2) The crash was caused by the pilots descending beneath 16,000ft contrary to the airline's instructions.
(3) The crew was not certain as to their position.
(4) The aircraft's radar would have depicted the mountain terrain ahead.
(5) The captain headed the aircraft toward cloud-covered high ground appearing to the pilot as an area of limited visibility or whiteout.

The Royal Commission of Inquiry conducted by Judge Peter Mahon opened on 7 July 1980 and sat for many months leading into the following year. The evidence presented by the airline was in keeping with Chippendale's report. The crew, somehow, should have detected the change in the co-ordinates leading to the alteration of track. No permission had been given by the airline for descent below 16,000ft on the approach to Mount Erebus and the fact that the captain had elected to do so was the primary cause of the accident. On this Air New Zealand was adamant. None of the airline's management or executive pilots were to be swayed from their opinion although subsequent witnesses were to prove a different story. It was revealed that before the flights began the company's Director of Flight Operations, at that time, Captain Keesing, had requested permission from the Civil Aviation Division (CAD) 'to descend to maintain at least 2,000ft terrain clearance'. Later Captain Gemmel, the commander of the first flight, wrote to CAD without Captain Keesing's knowledge stating that the minimum altitude for flight in the McMurdo area would be 16,000ft. This was eventually reduced to 6,000ft in the Sound. Captain Keesing was never apprised of the fact that his proposal had been superseded. In fact Captain Wilson, the route clearance unit briefing officer, testified that although 16,000ft and 6,000ft were established as the minimum safe altitudes in their respective areas, he admitted that he had given crews the impression that if the weather was clear they were free to depart from the assigned route and to descend to any level approved by ATC at McMurdo. Captain Wilson could reasonably be assumed by crews to carry the authority of the airline to permit such descents to lower altitudes. All flights from October 1977 had done so and had generally descended to between 1,500ft and 3,000ft. Several reports, including one in Air New Zealand's own in-house newspaper, had been published praising the trips and stated quite clearly that the aircraft had descended to heights of around 2,000ft. All of the executive pilots claimed not to have seen any of these accounts.

The file started by the airline to collect documents relevant to the accident was remarkably slim, and by coincidence seemed to contain only data in support of Air New Zealand's and Chippendale's case. Much material had apparently been destroyed. Captain Collins's ring-binder notebook, for example, recovered at the crash site, was returned to Mrs Collins with the pages missing. It is almost certain the detailed information on the briefing and the route were contained therein. It was only much later, after the Inquiry, that two policemen came forward to say that they had discovered the notebook at the crash site with the pages intact. What happened to these notes was never known. It also became known that a mysterious burglary had occurred at the home of Mrs Collins the previous year at about the time of the publication of Chippendale's interim report. Only a file of letters between Mrs Collins and her lawyers was taken.

In the course of the Inquiry the Chippendale report and Air New Zealand's stance came in for much criticism. The material of the airline briefing was considered inadequate and ill-informed and contained a number of mistakes. Only whiteout in snow shower conditions was considered and no warning of whiteout in other circumstances was covered. The fact that sector whiteout could occur in clear conditions and was not an unusual phenomenon in Antarctica but an everyday event was never conveyed to the crews. As Flight 901 flew south towards Erebus they faced a lowering cloud base meeting rising snow-covered ground, with the sun in the north shining from behind. By misfortune, the lower cliff face which might have given a visual reference was obscured by a fog bank which completed the ramp effect. A movie film of the accident site taken the day after the crash actually recorded the formation of this same fog bank. Light striking the snow-covered slope deflected upwards from the ice crystals and reflected back from the lower surface of the cloud. Such diffusion of light deceives the eye into seeing ahead a limitless vista and distant horizon. Crews were never warned of these dangerous effects. On Judge Mahon's own visit to the area he witnessed the same spectacle. On approaching the crash site in a helicopter the peak of Erebus could be seen sticking up above a low cloud layer, but at closer range the mountain simply disappeared from view leaving what appeared to be a distant horizon. The visual deception was completed by the amazing similarity between the entrance to Lewis Bay and the head of McMurdo Sound. Captain Vette, by a remarkable sequence of deductions, demonstrated quite clearly how the crew plus Peter Mulgrew had been tricked. The cliffs of Cape Bird to the right of Lewis Bay, for example, lay on a bearing similar to what would be expected of the cliffs at Cape Bernacchi to the right of McMurdo Sound. Although the Cape Bird cliffs were only one-third of the size they stood at almost exactly one-third of the distance from the DC-10's flightpath, and gave a similar impression of size.

The decision to excuse captains from a previous Antarctic visit before their own flight was severely criticised. Military crews interviewed at McMurdo were astounded to hear that Collins had never been to the area before. As the DC-10 had swept in a descending left turn to re-establish on the southbound track, passengers had photographed Beaufort Island on the right. If Collins had previous experience of the area, or had Mulgrew taken his seat just a few seconds earlier, the distinctive shape of the island would have been recognised and would have instantly indicated their position.

Liaison between Air New Zealand and the US ATC at McMurdo was also found to be inadequate. The area of descent designated by the company, the narrow 20-mile band to the south of Williams Field, was too close to the antenna for radar monitoring. The requirement, before the announced withdrawal of the radio beacon, for radar monitored descents down to 6,000ft in this sector was totally impractical. To monitor flights above 6,000ft within 20 miles required the radar antenna elevation to be reselected, prohibiting the monitoring of other flights on the approach below 6,000ft. One US witness described the sector as absurd. As a result descent within this sector was never accepted by US ATC, even after the announced withdrawal of the radio beacon when company requirements called only for descent to be co-ordinated with the controllers. The confusion resulted in crews assuming erroneously that radar cover above 6,000ft was not available anywhere.

The cockpit voice recorder (CVR) transcript submitted with the accident inspectors' original report was found wanting in a number of areas. Chippendale had edited the wording himself and some important comments had been omitted. Certain unintelligible phrases that had been discarded by experts in Washington were reincluded by Chippendale after a visit he made to Farnborough in England. 'Bit thick here, eh Bert' was one example which served to undermine the flight crew's position. No one in the crew was named Bert.

Chippendale's criticism of the crew's failure to use the weather radar in mapping mode for terrain avoidance was completely rejected by the Commission. Since the crew were flying in clear conditions, a fact supported by numerous passengers' photographs, it was doubtful if the weather radar had been switched on. If it had been in use it would more likely have been in weather mode. Statements from several radar experts, however, attested that in the very dry conditions of the Antarctic, drier even than the Sahara desert, radar returns from terrain covered in snow and ice would be negligible. The equipment could not under any circumstances be used in any mode for terrain avoidance purposes.

The airline was also criticised for not supplying charts with the actual route marked thereon. Without adequate charts Captain Collins had plotted his own route with subsequent results. The unbelieveable sequence of blunders by the navigation section, amounting to over 50 in all, was found to be the primary cause of the accident.

The report by the Royal Commission was submitted in April 1981, and completely exonerated the crew of any blame. 'In my opinion,' wrote Judge Mahon, 'neither Captain Collins nor F/O Cassin nor the flight engineers made any error which contributed to the disaster, and were not responsible for its occurrence.' Ten main events were deemed to have preceded the accident, the salient points being the lack of any chart showing a printed route, the change of the co-ordinate without the knowledge of the crew, and the effects of sector whiteout, eloquently described by Mahon as 'a malevolent trick of the polar light'. 'The dominant cause of the disaster', concluded the report, 'was the act of the airline in changing the computer track of the aircraft without telling the air crew.'

The recalcitrant stance taken by the airline management, and what appeared by Mahon to be a predetermined and organised conspiracy to divert the course of truth, prompted the judge to write that 'I am forced reluctantly to say that I had to listen to an orchestrated litany of lies'. Air New Zealand's Chief Executive, Morrie Davis, the man who had been behind the airline's stance, resigned and immediately challenged the report in New Zealand's Court of Appeal. In December 1981, by a majority judgement of three judges to two, the court upheld

Mahon's report on the cause of the crash but rejected his allegations of 'deliberate concealment and a concocted story'.

Mahon, upset by these findings, himself resigned from the New Zealand High Court to lodge his own counter-appeal with the Privy Council in London. Five leading British Law Lords sat to review the case. They found Mahon's report a 'brilliant and painstaking investigative work' and upheld his accusations that certain Air New Zealand personnel had been untruthful in the submission of their evidence. 'It is an understandable human weakness,' their Lordship's statement continued, 'on the part of individual members of the airline management to shrink from acknowledging, even to themselves, that something they had done or failed to do might have been the cause of so horrendous a disaster.' The Law Lords totally agreed with his findings as to the cause of the accident but 'very reluctantly felt compelled' to find that he had been excessive in his accusations of conspiracy. The witnesses in favour of the airline had never been presented at the Inquiry with charges of conspiracy and it was unfair to do so later. On this point of law, and on this alone, the Privy Council dismissed Mahon's appeal.

Above:
Beaufort Island as seen from a camera retrieved from the crashed aircraft. The photographs show how easy it is to mistake the dark island for a stretch of open water.

Below:
The wreckage on Mount Erebus. *Associated Press*

The Korean 747 Shoot-down   1983

The big Korean Air Lines jet flew on through the night on the journey to Seoul in South Korea. On the flight deck the navigator made preparations for the difficult North Pole route which lay ahead. With the magnetic compass unusable in polar regions and radio beacons unavailable, it was not an easy flight to navigate. Once above a certain latitude the compass would have to be set to gyro mode in which it would simply compensate for earth rotation. It would then have to be aligned with grid direction, the lines of which were overprinted on navigational charts. Astronomical navigation would be required and a periscopic sextant would be placed in position through an aperture in the flight deck roof. Observed altitudes of stars in groups of three would be compared at intervals with calculated details and from these readings positions would be accurately plotted. It was a time consuming process permitting only one star fix every 40min. The mental gymnastics involved in aligning the compass with grid north and preparing the sextant for star shots needed close concentration to avoid error. Working at night, in an alien time zone, it was easy to make mistakes.

Captain Kim Chang Kyu set his instruments according to his navigator's calculations and complied with his instructions. The captain was unable to monitor the flight without the aid of radio beacons and relied completely on the navigator to direct the navigation of the aircraft. The initial northbound route lay over the Arctic Ocean, remaining clear of Soviet airspace, and at first progress seemed satisfactory. After a period of time, however, an error by the navigator resulted in the aircraft going astray and the Korean flight was led 1,000 miles off course. In the first light of dawn the co-pilot caught sight of movement out of the corner of an eye and turned to see a fighter aircraft formating on the starboard side. On the interceptor's tail was the red star of the USSR. The fighter flew off before the captain could catch more than a glimpse, but he immediately began to flash the lights in recognition of being intercepted. Repeatedly he called on the emergency frequency in an attempt to establish contact, but to no avail. The Soviet fighter fell back behind the unsuspecting Korean Boeing 707 and for a moment the sky seemed clear. Suddenly, without further warning, the interceptor pilot took aim at the target and fired a burst from his cannons at the defenceless civilian aircraft. The shells sheared off the outer 15ft of the left wingtip of the 707 and shrapnel from the blast tore a hole in the forward fuselage. Two passengers were killed and 13 others injured. The cabin instantly depressurised and the captain immediately had to instigate an emergency descent from the 35,000ft cruising level to a lower altitude. Desperately Captain Kim searched for a suitable spot to land his striken aircraft and somehow managed to execute a wheels-up forced landing on a frozen lake near the town of Kem, 300 miles south of Murmansk. By an amazing piece of airmanship all but the two passengers killed by cannon fire were saved.

The Russian response to the international protests that followed the shooting-down of the unarmed 707 were predictable. The Korean aircraft, flying unannounced over the heavily fortified Kola Peninsula, was suspected of being a spy flight. The Soviets claimed they had tracked the intruder for more than two hours, their fighters had circled the 707 several times, and even warning tracers had been fired. The Korean crew, it was alleged, had failed to respond to all signals and as a result the order had been given to force down the offending aircraft. Subsequent examination of the wreckage and interrogation of the crew revealed no evidence whatsoever of complicity in a spying mission. It was established beyond all doubt that the 707 flight, on a direct routeing from Paris to Seoul, had experienced an unfortunate navigational error and that Soviet airspace had been penetrated quite accidentally. In due course the passengers and crew were returned to South Korea. An internal company inquiry completely exonerated the captain of any blame and he returned to flying duties, subsequently moving on to fly the Boeing 747. The Soviet Union remained unrepentant. It was made abundantly clear that in the future any aircraft violating their airspace in a manner similar to the Korean Air Lines flight on that fateful day of 20 April 1978 faced the same risk. The borders of the USSR were sacred!

Sadly, in that decade, the Soviets were not the only ones who had resorted to such drastic action. In 1970, an Alitalia DC-8 was struck by missiles fired from a Syrian fighter near Damascus. Fortunately it landed safely. In 1973 a Libyan Arab Airlines Boeing 727 strayed over occupied Sinai and was fired upon and forced down by Israeli interceptors. The aircraft was destroyed and 106 people aboard were killed. Like the Soviets, the Israelis insisted that the civil airliner had refused to comply with instructions. These incidents left little doubt in the minds of civilian pilots that the accidental penetration of forbidden airspace was not going to be taken lightly, and that the consequences of such action could be very serious indeed. There was every indication that any infringement in the future could lead to disaster.

On a wet and windy summer evening at New York's JFK Airport, passengers began checking-in for Tuesday's late night departure of Korean Air Lines flight KE007 to Kimpo Airport, Seoul, via Anchorage in Alaska. The date was 30 August 1983. The South Korean airline was handled in New York by American Airlines whose terminal building was also made available for their use. The passengers waiting to board Flight KE007, a Boeing 747-200B, registration HL7442, faced a long and tiring journey. The scheduled departure of near midnight out of Kennedy led to a seven-hour flight through the darkness with an arrival in Anchorage the next day. After a short transit the Boeing 747 would then proceed to Seoul, arriving, after another eight-hour journey under the stars, on the following morning. This routeing avoided overflying the airspace of the USSR, although it lay very close to its eastern border.

On the Anchorage-Seoul sector Flight 007 would cross the international date line and the travellers would jump one day ahead. The flight path of the Korean 747 from New York (local time GMT−4) to the transit at Anchorage (−8) and on to Seoul (+9) was along the shortest route, passing over the top of the earth on a course which lay close to the great circle track, the line joining the shortest distance on a globe between departure point and final destination. Such a northerly track traversed a large number of time zones. The time changes resulted in both sectors being flown through the hours of darkness, with each long leg beginning late at night, and was a tiring ordeal for the crews as well as the passengers.

The Korean Boeing 747 lifted off from JFK at about 00.20hrs local New York time (04.20hrs GMT) after a 30min delay and arrived at Anchorage at 03.30hrs local Alaskan time (11.30hrs GMT). Scheduled departure for Seoul was 04.20hrs local (12.20hrs GMT), which allowed just sufficient time for the passengers to stretch their legs in the spacious airport transit lounge and to browse through the many shops. As the weary travellers relaxed in the terminal building the 747 was cleaned, replenished and refuelled for the next stage of the journey. The crew which had operated from New York disembarked for a period of rest in Anchorage while a fresh crew of three flight deck personnel and 20 cabin staff — seven stewards and 13 stewardesses — prepared to board the flight. About 15min after KE007's arrival another Korean Air Lines flight, KE015 from Los Angeles, also bound for Seoul, parked at the terminal and its passengers swelled the numbers mingling in the transit area.

Flight KE007 was now to be commanded by Captain Chun Byung In who, with his co-pilot, First Officer (F/O) Sohn Dong Hui, and Flight Engineer (F/E) Kim Eui Dong checked the flight details in the operations room before proceeding to the flight deck. Continental Airlines prepared the paperwork, the flight plan of which had been telexed earlier from Los Angeles. All the information was carefully examined and the fuel uplift approved by Captain Chun. Only one item on the notices to airmen (Notams) was of any relevance and that was the unserviceability of the VOR radio beacon at Anchorage. The flight time, however, required some attention. The scheduled Anchorage to Kimpo journey time was 8hr 20min, but because of favourable forecast winds the computer flight plans indicated a flight time of only 7hr 35min. A quick mental calculation in round figures revealed the problem. Departing from the gate at 04.20hrs Anchorage local time gave an airborne time of about 04.30hrs which, when the eight hours in flight were added, resulted in an estimated arrival time at Seoul of 05.30hrs local Korean time, 30min earlier than planned. South Korea, in keeping with other countries such as Japan and Australia, had a policy of closing their airports at night. No jet air traffic movements were permitted. Kimpo airport was not due to open until 06.00hrs local, so the simplest solution was to delay departure of KE007 by 30min.

Meanwhile, Captain Chun and his crew boarded the aircraft and commenced their pre-flight checks. Amongst the flight crew's tasks was programming of the inertial navigation system (INS) which would navigate the Boeing 747 on the next leg of the flight to Kimpo airport. In the last decade civil aviation had witnessed the demise of the navigator and all major intercontinental routes were flown using such electronic navigation devices, or similar equipment. INS is a spin-off from American space projects and is a self-contained airborne unit completely independent of ground-based navigation aids. The system consists of a gyro-stabilised platform on which are placed accelerometers which detect movement of the aircraft in all directions. These movements are processed by the computer and used to compile navigational data. Since the INS always knows where it is, and the computer programme contains details of earth shape and movement and direction of true north, the system is able continuously to display actual position as the flight progresses as well as navigate the aircraft to a distant point. Other navigational data such as speed, distance, track and wind are also available. In addition signals from the INS supply stable reference datums for certain flight instruments and for autopilot function. With autopilot engagement, INS mode can then be selected for automatic navigation of the flight.

On the arrival of KE007 in Anchorage, the INS computers on board would retain in their memory the position of the aircraft at the terminal ramp and would even compensate for earth movement during the time on the ground. After long journeys, however, small errors can creep in, and, extremely accurate although the system may be, the computer may 'think' it is a mile or two away from its actual position. This is resolved by selecting the INS to an 'alignment' mode and by informing the machine exactly where it is. The precise latitude and longitude of the aircraft's position, given to the nearest tenth of an angular minute in the Anchorage Airport booklet (eg Anchorage 61° 10.7'N, 149° 59.2'W) is inserted into the computer via a numbered keyboard. The INS then compares the actual position inserted by the pilot with where it 'thinks' it is and corrects itself accordingly. It also works the other way, and can detect a ramp position incorrectly inserted by the pilot — known in the trade as 'finger trouble' — by illumination of a red warning light. If the error is significant the pilot can then recheck his figures and reset the correct position. In the align mode, by the simple expediency of sensing earth rotation, the computer recalculates true north from the 'corrected' position and adjusts for its previous errors in an attempt to eliminate them on the next flight. The alignment process can take up to 13min and during this period the aircraft must not be moved. A green 'ready' light illuminates when the process is completed and the selector must then be positioned to navigation mode before the aircraft moves from the ramp. Checking and setting of the INS to align, and inserting the ramp positions, is normally the function of the co-pilot. All big jets such as the Boeing 747 normally carry three completely independent inertial navigation systems, so the insertion of the ramp position for the alignment process has to be undertaken three times. If only two INSs were on board — like someone wearing two watches — and one malfunctioned, it would be impossible to tell which was operating normally. With three systems it can be reasonably assumed that with any two in agreement the third one must be incorrect. Whilst navigating the systems are electronically mixed, and a compromise position is obtained from the mid-latitude and mid-longitude of the three INSs. By such an arrangement any malfunction of one system does not give a large navigational error.

Once selected to navigation mode the INS, via an engaged autopilot can automatically navigate the flight in any desired direction, but first has to be programmed with the route. The pilots insert the aircraft routeing into the computer by keying into the INS the exact latitude and longitude co-ordinates of positions along the way. Up to nine such INS locations, known as waypoints, can

200

be inserted at one time with the sequence being updated as the aircraft progresses. A remote button can be pressed on each set to allow simultaneous loading of waypoints to all three systems. The INS waypoint may represent the position of a radio beacon site or a compulsory reporting point with defined latitude and longitude, which over the ocean will simply be a convenient imaginary mark on a chart. The first significant waypoint on KE007's route was the position of the VOR radio beacon at the hamlet of Bethel, 345nm away, on the inlet of the Kuskokwin River on the west Alaskan Coast. The INS has been proven in service to be an extremely accurate device and an American study has shown that problems associated with pilot insertion of incorrect data occur only once in every 19,600 flights. The INS had also been demonstrated to be remarkably reliable and on no occasion has a triple failure of the system been recorded. INS is a most impressive navigational tool.

With the pre-flight checks completed and the aircraft prepared for service the passengers were finally called from the transit lounge for the delayed departure. As the travellers filed aboard, 'clearance delivery' was called on the radio for the departure routeing. KE007 was cleared for departure from runway 32 and the crew checked the details from the airport booklet: climb on runway heading as rapidly as practicable to 400ft then turn left on to track 300° magnetic and expect vectors to assigned route or depicted fix. Climb through 3,000ft as rapidly as possible. The initial routeing would direct Flight 007 on to airway Victor 319/Jet 501 to Bethel, or on a direct routeing assigned by radar to any point on the way, and then on route Romeo 20 (R20) over the Bering Sea along the line of the Aleutian Island chain. The route would then pass over the north Pacific Ocean and south of the Sea of Okhotsk, skirting just outside the edge of airspace controlled by the Tilichiki and Petropavlovsk-Kamchatskiy Soviet air traffic control centres, on through the northern Japanese airways network and across the Sea of Japan to South Korea. The total flight distance from Anchorage to Seoul was 3,566nm.

Once aboard, the 246 passengers, mostly Korean, Taiwanese and Japanese, including some Korean Air Lines crew returning to base, settled down for the long trip ahead. Amongst the sprinkling of other nationalities on the flight, including American, Canadian and British, was one gentleman of note, Congressman Lawrence McDonnald, national chairman of the right wing conservative John Birch Society. Almost all the passengers had commenced their journey in New York. The crew of 23 on KE007 brought the total head count on the Boeing 747 to 269 people. With the boarding completed the doors were shut and the engines were started in sequence. KE007 was cleared to push back and taxi to runway 32 and, with the start checks completed, the aircraft was soon on its way.

The big jet took-off from runway 32 into the night sky at exactly 05.00hrs local Anchorage time (13.00hrs GMT — now used for convenience throughout) and soon turned left on to the assigned heading of 300°. One minute after becoming airborne, as the 747 climbed quickly as per instructions, F/O Sohn switched the radio from 'tower' to 118.6MHz and called 'departure '. The radar controller replied with further instructions.

13.01:12 Departure Control R/T: 'Korean Air 007 heavy, Anchorage departure, radar contact. Climb and maintain level three one zero. Turn left heading two two zero.'

F/O Sohn R/T: 'Roger, two two zero, climb and maintain level three one zero, Roger.'

Captain Chun flew the aircraft on to heading 220° and the big jet banked left in response. The recleared level of 31,000ft was set in the autopilot capture window. Passing 3,000ft the aircraft was then accelerated for flap retraction and the speed increased to normal climb requirement. The autopilot was then engaged in heading mode with the heading cursor set on the magnetic compass to 220°.

Departure Control R/T: 'Korean Air 007 heavy, proceed direct Bethel when able.'

F/O Sohn R/T: 'Roger, uh, proceed direct to Bethel, Roger.'

Departure Control R/T: 'Korean Air 007 heavy, contact Anchorage Centre one two five point seven. Good day.'

F/O Sohn R/T: 'Good day.'

Bethel lay over 300 miles to the west with KE007 well outside the 200nm range of the VOR radio beacon at that position, so the INS was simply programmed for a direct routeing. Captain Chun touched zero (always the present position) and the waypoint number for Bethel on the INS keyboard, then pressed the 'insert' key. Immediately the direct track details from the actual position to Bethel were obtained. Chun turned the large 747 in the direction of Bethel using the button-sized autopilot heading control knob, then, it is assumed, switched the autopilot mode selector from heading to INS. The INS would now be coupled to the autopilot and the navigation system would, in that case, automatically fly KE007, allowing for changes in wind, on a direct track to Bethel. F/O Sohn established contact on 125.7MHz with the Anchorage en route controller who supervised the progress of KE007 via Kenai Radar Station situated southwest of Anchorage. Kenai radar observed the Korean 747 climb just north of airway Jet 501 and settle in the cruise at level 310. The aircraft then passed about six miles north of Sparrevohn VOR radio beacon — the sort of deviation a controller might expect from an aircraft cleared on a direct routeing — before moving out of the station's 175nm range. KE007 cruised on towards Bethel, remaining in the darkness as it flew westbound away from the rising sun.

At Bethel, F/O Sohn reported KE007's position as overhead, and passed an estimated time for Nabie, the first of the imaginary position reporting points out over the Bering Sea. The other oceanic reporting points were Neeva, Nippi and Nokka, leading onto the coast of Japan. The INS is sufficiently accurate to guide an aircraft over a position, but not always with the precision to pinpoint exactly a radio beacon location. On the instrument panel, therefore, the needles of the radio magnetic indicator pointing to a beacon usually fall to the right or the left during passage. As KE007 passed by Bethel, the needles fell to the left, but not as rapidly as they should have done had the aircraft been close to the VOR radio beacon. In fact, the needles fell quite slowly, indicating the aircraft to be flying some miles north of Bethel, an event which may have been overlooked by the crew at this busy stage of the flight. By now KE007 was outside civil radar cover, but an American military radar station at King Salmon, 220 miles southwest of Anchorage, spotted the 747 flying 12 miles to the north of Bethel. The military installation was part of a chain of radar stations monitoring the approaches to

Alaska from the east and was not concerned with flight planned civil traffic departing the region to the west. At that time there was little or no liaison between civil and military aviation organisations, a situation which was not satisfactory but was understandable. The capabilities of military radar installations are classified and any intervention on behalf of civil flights could compromise security. KE007, for all the King Salmon military observer knew, was simply following civil air traffic control instructions.

On the 747 flight deck all seemed well as the INS appeared automatically to guide the aircraft along the 239° magnetic track towards Nabie, 312 miles to the west. But the situation was far from well. The 747 had not only passed 12 miles north of Bethel but was still diverging from course. The track, in fact, led to the north of Romeo 20, on a route to danger which passed over the sensitive Kamchatka Peninsula in Soviet territory. The area was under constant military radar surveillance, and that night Soviet vigilance was to be tested. Meanwhile, the Korean flight crew, apparently unaware of their predicament, settled down to the familiar procedure of monitoring progress and passing position reports where required. At one minute before each waypoint an amber 'alert' light illuminated on the INS (it also functioned in the same manner when passing abeam a position up to 200nm away) to warn the crew of approach to the next position. The INS could then be checked as it turned on to the next track, and the times recorded for transmission to the relevant air traffic control. At about 14.30hrs, KE007 passed abeam Nabie and F/O Sohn attempted contact with Anchorage on very high frequency (VHF) radio. Although outside normal VHF cover, an

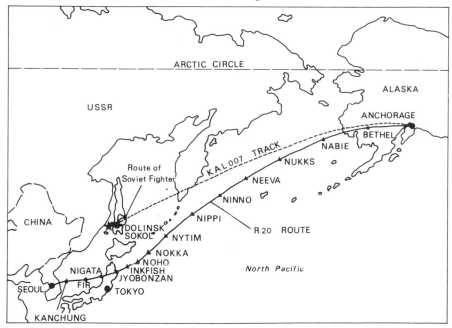

Above:
The intended, R20, route and actual KAL007 track.

203

automatic VHF link along the Aleutian chain of islands relayed the calls. Korean Air 007, however, was not at Nabie, but was by now 40 miles north of track and was just outside VHF range of the nearest relay station to the south. Fortunately the other Korean Air Lines flight at Anchorage, KE015 from Los Angeles to Seoul, which had departed 20 minutes later, heard the call and relayed the details. KE007's estimate for Neeva was transmitted at 16.00hrs. No indication was given of any problem and every impression was imparted that the Romeo 20 route was being flown. As a check KE007 established high frequency (HF) long range communications with Anchorage at 14.44hrs, but at Neeva, at 15.58hrs, F/O Sohn again relayed his position on VHF via KE015. The estimate given for Nippi, 561 miles away, was 17.08hrs. At Nippi, control would change from Anchorage to Tokyo and the instruction was given to call Tokyo on en route radio at that point. Passing Neeva, KE007 had strayed 150 miles off track.

At 16.06hrs F/O Sohn requested clearance to climb to 33,000ft, and Anchorage control relayed their approval. The big jet ascended 2,000ft to the next height and cruised on towards danger at level 330. If the crew were aware of their actual position they remained remarkably calm, for just under 400 miles away lay the coast of the hostile Kamchatka Peninsula. The Korean 747 was, by now, flying unannounced in an area supervised by Soviet civil controllers but was still well outside Russian airspace. It did not, however, proceed unnoticed. Already Soviet military observers were tracking KE007 on their radar screens.

Captain Chun would also likely have been using radar, in this case to pick up weather, with the scanner tilted down about one degree to detect isolated thunder clouds. The moon, by now, had risen and shone almost half full in the southeast giving some light to the sky, so he may also have been able to see cloud formations. Some ground returns would have been visible on the weather radar screen, but low cloud was expected on the route and may have confused the picture. Had Captain Chun found the need to select mapping mode, it is likely that a clearer outline of the Kamchatka Peninsula would at this stage have been coming into view.

The Soviet radar operators, observing KE007, now noticed another unidentified target appear on the screens about 75 miles ahead of the 747. As the Russians may have suspected, this was a US Air Force RC-135, an adapted Boeing 707 on an intelligence gathering mission. Its presence in the area was not unexpected as such aircraft patrolled the region on up to 20 days per month and sometimes a relay of aircraft remained on station for 24 hours a day. The behaviour of the RC-135 did not give cause for concern to the Soviets, but by now they were becoming extremely disturbed by the approach of the unidentified Korean 747. F/O Sohn's radio reports may have conveyed an air of calm on KE007's flight deck, but on the ground there was a great deal of consternation. As the 747 approached the Kamchatka Peninsula with its early warning radar systems, missile testing sites and the port facilities at Petropavlovsk, a base for nuclear armed submarines, the Russians could no longer exercise restraint. Six MiG-23 all-weather fighters were scrambled. KE007 took 24 minutes to overfly the Kamchatka Peninsula with the MiGs in hot pursuit, but in spite of being tracked by radar the interceptors were unable to locate their prey. As the Korean flight left Russian airspace over the Sea of Okhotsk the MiGs broke off the chase and ignominiously returned to their stations. The Soviet defence system had been found somewhat lacking.

KE007 passed abeam Nippi at 17.07hrs and should at that moment have been entering Japanese controlled airspace. The aircraft was, in fact, 185 miles off track. F/O Sohn reported his position as Nippi to Tokyo on HF radio and transmitted the estimate for Nokka, another imaginary reporting point lying 660 miles away, about 100 miles south of the Kuril Islands.

17.09 F/O Sohn HF R/T: 'Korean Air 007, over Nippi one seven zero seven, level three three zero, estimate Nokka at one eight two six.'

The Korean 747 cruised over the Sea of Okhotsk, now back in international airspace, and continued on a heading towards the southern tip of Sakhalin Island. The island was another Soviet hot spot, controlling the approach to the Sea of Japan and the important naval base of Vladivostok. On the south coast of Sakhalin, by the Soya or La Perouse Strait, is situated the strategic naval establishment of Koraskov, surrounded by a number of airforce bomber and fighter stations and a major missile site. It was a highly sensitive area, second only to the submarine base at Petropavlovsk.

The embarrassment felt by the Soviet defence commander at the failure to intercept the unidentified aircraft could not go unchecked, and every effort was being made to prevent the escape of the intruder. As far as the Russians were concerned the aircraft was probably a US RC-135, although its speed seemed too fast. Perhaps it was a converted Boeing 747 — an E-4A or E-4B intelligence gathering aircraft. It might even have been a civilian aircraft pretending to be off track; their own airline Aeroflot was suspected of indulging in such practices. Whatever it was it could only be on a spying mission and it had to be intercepted and forced to land. Soviet fighters on the Kuril Islands were scrambled to block any escape route to the south. Japanese and American intelligence officers, unaware of the situation, simply assumed the Russians were involved in an air defence exercise. On Sakhalin Island, at the Dolinsk-Sokol air base, crews were briefed and more fighters were placed on alert. As the 747 continued on its course, they too were scrambled to intercept.

Time was now running short for the Russians. The south of Sakhalin Island is at its widest only 80 miles across and the 747 would traverse Soviet airspace in less than 10min. Just a few miles southwest of the island at the far side of the Soya Strait, lay the northern Japanese Island of Hokkaido and the forbidden area of Japanese airspace. If KE007 was allowed to proceed beyond the shores of Sakhalin Island it could not be followed. It had to be intercepted before that point.

The Soviet fighters climbed rapidly from their Sakhalin base to 33,000ft, and under radar control flew out across the Sea of Okhotsk to meet the intruder. With radar guidance the interceptors turned behind the approaching 747 and took up positions on its tail. Japanese military radar stations observed the scene and saw what appeared to be a Russian transport aircraft being escorted by some fighters. Electronic ears were also listening to the activity and air-to-ground transmissions between the fighters and their control centres were recorded. Three fighters were apparently involved, with callsigns 805, 163 and 121, conducting communications between control centres codenamed Deputat, Karnaval and Trikotazh.

Ground-to-air radio instructions between these stations and the interceptors were also believed to be recorded, but were not released. Aircraft 805, identified as a Sukhoi Su-15, and ground control station Deputat, played dominant roles in the events which were about to unfold.

After being radar vectored onto the intruder's tail, 805 picked up the target on his own radar and moved in for a closer look. KE007 was flying on course 240° at 33,000ft and at a ground speed of about 500kt. At just before 18.06hrs, as the Su-15 closed the gap on the 747, fighter pilot 805 made a jubiliant call to his ground control station.

18.05:56 805 R/T: 'I see it!'

Deputat responded with further directions to which the Su-15 pilot replied:

18.08:00: '805, Roger. Understood. I'm flying behind.'

18.08:06: '805, on course two six zero, Understood.'

The Su-15 fighter turned slightly right and pulled almost abreast of the 747's right side, but remained some way off. Mistakenly KE007 was reported as changing course.

18.09:00: '805, yes, it has turned. The target is 80° to my left.'

18.10:35: '805, course two two zero.'

18.11:20: '805, 8,000m, Roger.'

18.12:10: '805, I see it visually and on radar.'

The moon was still visible, although less than half full, but should have given sufficient light to identify the aircraft in the clear sky. It must have been obvious by now that the aircraft was not an RC-135, but it perhaps could have been identified as an E-4B. If KE007 was recognised by the Su-15 pilot as a civilian aircraft the information was never transmitted to ground control. It was not 805's responsibility to identify the target; he was simply obeying instructions. 805 then activated an identification procedure known as IFF — identification friend or foe — but with inevitable results. Soviet systems are obviously not compatible with other equipment.

18.13:26: '805, the target isn't responding to IFF.'

On the flight deck of the 747, the Korean crew seemed oblivious of the activity and were more interested in climbing to the next suitable flight level. The co-pilot radioed Tokyo on HF for approval.

F/O Sohn HF R/T: 'Korean Air 007, requesting three five zero.'

Tokyo HF R/T: 'Roger, standby. Call back.'

Meanwhile the Su-15 pilot continued with his transmissions.

18.15:08: '805, the target's course is still the same . . . two four zero.'

18.18:07: '805, I see it.'

18.18:34: '805, the ANO are burning. The light is flashing.'

The Su-15 pilot confirmed the navigation lights (as stipulated in the Air Navigation Order) were lit and that the flashing red anti-collision light was operating.

18.19:02: '805, I am closing on the target.'

18.19:08: '805, they do not see me.'

18.20:08: '805, fiddlesticks. I'm going. That is, my ZG is lit.'

805 verified that his ZG indicator was illuminated, which confirmed his radar guided missiles were locked on target. He remained at lock-on for only a short

period then called back, perhaps following further instructions.

18.20:30: '805, I am turning lock-on off and I'm approaching the target.'

At about this moment KE007 received a call from Tokyo with approval to climb.

Tokyo HF R/T: 'Korean Air 007, clearance. Tokyo ATC clears Korean Air 007 climb and maintain level three five zero.'

F/O Sohn HF R/T: 'Roger, Korean Air 007, climb and maintain flight level three five zero, leaving three three zero this time.'

There was no indication of anything untoward occurring, although at this point the 747 was 365 miles off track. Captain Chun selected vertical speed mode on the autopilot and began the ascent to 35,000ft, with the aircraft slowing slightly as it did so. As the climb clearance was being received the Su-15 pilot fired 120 rounds from his cannons in four bursts, ostensibly as a warning to 007. If any shells did strike the aircraft the incident was not reported. The Korean crew, busy with the climb procedure, did not appear to see any tracers, if that was 805's intentions.

18.21:35: '805, the target's light is blinking. I have approached the target to a distance of about two kilometres.'

The Su-15 pilot, increasing speed to close the gap, was caught off-balance with the 747 climb manoeuvre.

18.22:02: '805, the target is decreasing speed.'

805 inadvertently overtook the intruder.

18.22:17: '805, I am going around it. I'm already moving in front of the target.'

18.22:23: '805, I have increased speed.'

18.22:29: '805, no. It is decreasing speed.'

18.22:42: '805, it should have been earlier. How can I chase it? I am already abeam of the target.'

18.22:55: '805, now I have to fall back a bit from the target.'

KE007 now called Tokyo confirming that 35,000ft had been reached.

F/O Sohn HF R/T: 'Tokyo Radio, Korean Air 007, level three five zero.'

The Su-15 pilot received more instructions from Deputat to which there was a constant stream of replies.

18.23:18: '805, from me it is located 70° to the left.'

18.23:37: '805, I am dropping back. Now I will try rockets.'

18.24:22: '805, Roger. I am in lock-on.'

18.25:11: '805, I am closing on the target. Am in lock-on. Distance to target is eight kilometres.'

18.25:46: '805, ZG.'

The Sukhoi Su-15 pilot fired his missiles.

18.26:20: '805, I have executed the launch.'

The two AA-3 'Anab' rockets streaked toward the 747 and a couple of seconds later struck the tail of the target and probably an inboard engine. The big airliner instantly broke up and spiralled towards the ground. The interceptor pilot watched as the 747 exploded in the darkness, then matter-of-factly called ground control.

18.26:22: '805, the target is destroyed.'

18.26:27: '805, I am breaking off the attack.'

In the 747 cockpit, the flight crew, surviving the initial strike, were taken completely by surprise. Desperately the co-pilot transmitted a few garbled words to Tokyo Control.

18.27 F/O Sohn HF R/T: 'Korean Air 007 . . . all engines . . . rapid decompression . . . one zero one . . . two delta. . . .'

Nothing more was heard from KE007. The time was 18.28hrs GMT on 31 August, 03.28hrs local Japanese and Korean time on 1 September. Japanese military radar operators witnessed the 747 spiralling from the sky and a few minutes later at about 18.30hrs saw the target strike the sea off the west coast of Sakhalin Island, just outside Soviet territorial waters. All 269 aboard the Korean 747 perished.

News of the shoot-down broke upon a stunned world and provoked an international outcry. Angry retorts from leaders of the free world denounced the Soviet Union and demands were made for retaliatory action. The Canadians were the first to stop Aeroflot flights into their country and other nations followed suit. President Reagan closed Aeroflot's offices in New York and Washington and a worldwide 60-day ban was called on all flights to Russia. The Soviet reaction to the fury was predictable. At first they refused to admit the act, but after a few days eventually confessed to the shooting down of KE007 in mistake for a US RC-135 spy flight. As far as they were concerned, however, the intruder was on a spying mission and they had every justification in shooting it from the skies. An impressive propaganda exercise was mounted by the Soviets to support their claim and they openly condemned the United States for their 'deliberately-planned provocations'. In an unprecedented move, the chief of the Soviet general staff, Marshal Nikolai Ogarkov, held a press conference in Moscow to present their case. A large chart displayed how, in the Soviet opinion, the Korean 747 had rendezvoused with the US RC-135 before proceeding deliberately to violate Russian airspace. Marshal Ogarkov, giving almost a repeat account of the shooting down of the Korean 707 south of Murmansk in 1978, claimed that the 747 had been tracked by military radar for several hours, that the aircraft was flying without lights, that the crew did not respond to attempts at contact and that warning shots were ignored. 'There was no doubt that it was a reconnaissance plane' and 'the aircraft's destruction and the loss of life should be blamed on the US'. In the words of the *New York Times*, Marshal Ogarkov gave a 'spellbinding performance'. The incident soured US-Soviet relations and injured efforts at detente. One year later, however, both sides had greatly modified their stance: Washington no longer maintained that the shooting was a deliberate and callous act and Moscow openly admitted that the shooting was a grave error.

But why was KE007 so far off track? Was the Korean 747 on a spying mission? The Russians made accusations that KE007 was carrying sophisticated electronic eavesdropping equipment, but this is difficult to accept. How this aircraft was supposed to be serviced by other airlines at distant stations without the mechanics' knowledge that such devices were on board was not explained. The array of antennas required would have been obvious to all. The so-called 'rendezvous' with the RC-135 can only be dismissed as coincidental. At that time these aircraft were in the region almost every day and at no point were the two flights closer than 75 miles. The RC-135 was back at its Shemya base at the end of the Aleutian Island chain about one hour before the shoot-down of the 747. The

Soviets, it was suspected, were preparing to test a new weapon that night, but why should the Koreans take such risks when intelligence-gathering aircraft could patrol effectively at a safe distance and spy satellites could effectively gather intelligence from afar? The Soviet claims of spying are unlikely.

There were plenty of other rumours, however, some originating in the West, that the 747 intrusion had an equally sinister purpose. It was suggested by some that the Korean 747 carried no eavesdropping equipment, but its violation of Russian airspace was arranged simply to provoke a response from the Soviets which could be monitored by the West. It was claimed that the delay of KE007 from Anchorage was arranged in order to maintain VHF radio contact with KE015 who transmitted fake radio calls, that a rendezvous with the RC-135 did take place, and that the intrusion was timed to coincide with the passage of the US Ferret spy satellite and the space shuttle *Challenger*, which had been launched the day before. There is no doubt that the Soviet activity resulting from the Korean 747's demise provided a windfall for US intelligence, but the suggestion that the infringement was planned is totally unfounded. The insinuations that a civilian airliner pretended to be off track either to gather intelligence or to provoke a defence response for monitoring purposes are both based on the premise that the Russians would not shoot down a civilian aircraft. Yet only five years earlier the Soviets had demonstrated quite clearly that they would not tolerate infringement of their borders. If any air carrier was aware of the consequences of inadvertent violation of Russian airspace, it was Korean Air Lines. The accusations of spying in this manner are also unlikely.

The same can be said for the charges of deliberate corner-cutting. All pilots, at some time, take short cuts, perhaps following the central line of a twisting airway, but an attempt to cut a corner on this scale is unimaginable. Even had Captain Chun deliberately tried to follow the shortest distance along the great circle track from Anchorage to Seoul and had avoided the Soviet interceptors, how was he going to fool the Japanese authorities? Sending 'fake' position reports on HF radio over remote areas is one thing, but trying the same trick on VHF through a congested radar controlled airways system is another matter. In most radar controlled areas position reports are not even required, and after initial contact aircraft progress is simply monitored by radar. Since KE007 would have passed to the north of Japan instead of approaching the east coast, it would have been well outside VHF range of the airway sector controller situated 350 miles to the south. Even if F/O Sohn had been able to establish contact, the controller would hardly accept position reports from a flight he couldn't even see on his radar screen. Yet this is exactly what has been suggested in some quarters. The Japanese would more than likely have scrambled their own fighters if the unidentified 747 had appeared in their airspace in the north, and with the failure of KE007 to appear as planned on the airway system they would immediately have alerted the search and rescue services. Even had Captain Chun managed to outwit the Japanese defence systems he would have had to cross a north bound airway unannounced, pass through a high altitude training area, and breach the Korean air identification zone before reaching Kimpo Airport.

The suggested intention of Captain Chun's corner-cutting was, of course, to save on costs, but this argument does not stand up to scrutiny. Simulated flights of

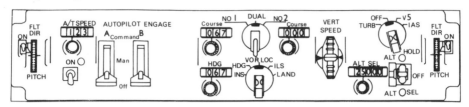

both the Romeo 20 and great circle routes have shown that by going direct the total distances would have been reduced by only 178 miles and the flight time by 20 minutes. Since the departure time from Anchorage was delayed by 30 minutes to co-ordinate KE007's arrival at Seoul with Kimpo Airport's opening time of 06.00hrs, what would be the point in making up time just to waste fuel holding over destination? In this respect alone the accusations of corner-cutting seem inconceivable.

Other suggestions as to the demise of KE007 concern incorrect INS and autopilot procedures. One theory is that after Captain Chun turned the aircraft towards Bethel using the autopilot heading control knob, he omitted to switch to INS and the selector was left in the magnetic heading mode. The selection of INS is accomplished by turning an autopilot switch anti-clockwise one notch from heading to INS. In front of both pilots green lights then illuminate on a flight mode annunciator to indicate that the navigation track has been captured. The insinuation that Captain Chun forgot to rotate this mode selector one division is unlikely in this case although there are some arguments in favour of the theory. There is no doubt that similar incidents had occurred. As recently as November 1985, a Japan Air Lines Boeing 747, en route from Tokyo to Moscow, strayed 60 miles off course for that very reason and flew close to the same spot near Sakhalin Island where KE007 went down. The Captain had switched from INS to magnetic heading to fly around a thundercloud but had forgotten to reselect INS mode after negotiating the weather. Strong winds had blown the 747 westward. The crew discovered the error when completing a regular INS reporting point check and after repeated attempts managed to contact Khabarovsk air traffic control centre. In the meantime two Soviet MiG fighters were scrambled. Fortunately the situation was sorted out in time and the 747 allowed to continue unhindered. The argument in Captain Chun's case is that after establishing on the airway centre line from Anchorage to Bethel he omitted to select INS mode on the autopilot and continued on in magnetic heading mode. The wind at the time resulted in a left drift of only one degree so a magnetic heading of 246°, corresponding to the initial magnetic airway track, would lead the aircraft close to Bethel. If the aircraft then continued on a magnetic heading of 246° following an extension of the straight Anchorage to Bethel airway, it has been demonstrated that the course flown by KE007, that is, taking account of the wind and magnetic variation (the difference between true and magnetic north), leads to the shoot-down point over Sakhalin Island. Another convincing fact is that the course of 240° as reported by the Su-15 fighter pilot corresponds with KE007's. The wind aloft in that area

Above:
Korean Air Lines 747.
Via John Stroud

Left:
INS control/Display Unit.

Below left:
747 flight deck.   *Boeing*

211

resulted in a left drift of 5° and on a heading of 246° the 747 would fly a track of 241°. This implies, however, that the flight crew must have sat for five hours with the selector in heading mode and with no illumination of the green navigation capture lights without anyone noticing. It also indicates that not one crew member checked any of the INS sets at any reporting point. It is almost impossible to imagine that an entire crew could be this irresponsible and on these facts alone this theory is unlikely. There are, however, more compelling reasons to indicate that INS *was* selected.

On initial departure KE007 was cleared to 'proceed direct to Bethel when able'. This instruction means, quite simply, that the flight was cleared from the aircraft's present position direct to the assigned point. There was no requirement for the Korean flight to proceed along the airway. The addition of 'when able' normally applies to non-INS equipped aircraft which can proceed direct to a distant VOR radio beacon only when within range of the signal. The skies of Alaska are full of such aircraft and the controller said 'when able' out of habit, even to the 747 which could proceed direct at the push of a button. It has been suggested, however, that 'when able' was added to the clearance because of mountains to the north and west and that it was necessary for Captain Chun to fly the airway route between Anchorage and Bethel to avoid the high ground. At the time Captain Chun received the instruction to proceed direct to Bethel, KE007 was heading southwest over the low lying coastal area, well clear of the 5,500ft mountainous terrain lying 25 miles to the northwest and known locally as the Sleeping Lady. To the west the high ground stood at 5,000ft and lay 45 miles distant. Proceeding direct to Bethel at that moment would not infringe minimum safe altitudes. The only way the 747 could proceed on a straight line from its present position north of the airway to the out-of-range VOR radio beacon at Bethel was by use of the INS, which is standard navigating procedure, and it is most likely that Captain Chun would have programmed the INS and autopilot to fly direct to Bethel. This is the strongest argument in favour of the INS being engaged and narrows the fault to a programming error or to unit failures.

Whatever the error or fault in the INS the problem must have occurred before departure, or very soon afterwards, for almost immediately the flight began to go wrong. The insertion of incorrect waypoints can be discounted for it is impossible to imagine them all being wrong, but it is possible that the initial ramp position at Anchorage was incorrectly inserted. Analysis of data has shown that a 10° easterly error in longitude, ie 61° 10.7′ N; 139° 59.2′ W instead of the correct 61° 10.7′ N; 149° 59.2′ W results in a flight path which also leads to the shoot-down point. The theory is also in keeping with the 240° course indicated by the Su-15 pilot. Although the INS position over the earth was incorrect, the directions of waypoints relative to each other would be retained correctly in the INS memory. The true track between Nippi and Nokka was 229°, and with the addition of the magnetic variation in the region of 9°, a magnetic course of 238° results. To achieve the ramp position error, however, the pilot inserting the data would have to key the longitude incorrectly three times as he separately loaded each INS. If this alleged error did occur it seems strange that one digit should be incorrectly entered in such a way and that it was not picked up by the crew. The entry of 139° 59.2′W instead of 149° 59.2′W appears an unlikely discrepancy and one that would be sufficiently outstanding to be visually detectable. The transposing of two numbers is more likely ie 159° 49.2′W instead of 149′ 59.2′W, thus displaying the same digits but in a slightly different sequence. Such a mistake is also more difficult to detect, but in this case results in a 10° (to be exact 9.8°) westerly error. Assuming an error did occur, however, three red warning lights would then be illuminated. The incorrect ramp positions would then have to be inserted once more into each system before the INS would accept the error and the red warning light would extinguish. Had the error been of similar significance in latitude it would not have been accepted, for the INS can detect peripheral velocity of the earth as a cross check. The aircraft's longitude, however, is held only in the memory with no means of cross-check, and a second insertion of an incorrect longitude can override the data. Each flight crew member, though, would then be required independently to check the ramp position at a later stage and, if the theory is correct, would have to fail to recognise the error. Once the aircraft moved off, there would be no way of re-checking the ramp position. Incidents of this nature, although extremely rare, had occurred in the past. Captain Chun and his crew boarded fresh in Anchorage, although 'fresh' on that route is only relative bearing in mind the large time changes and journeys through the night. The flight crew, however, were not in a rush, gave no indication of being anything but alert, and such a sequence of errors, although possible, does not seem probable.

So what of INS failure? A simultaneous failure of all three systems is unknown although two have been known to break down on the same flight! INS has also been recorded as suffering from a number of isolated faults. Could an insidious series of events have resulted in the 747 going astray? As has been demonstrated in the 1979 Chicago incident involving the DC-10 engine pylon detachment, if something can happen in aviation, no matter how improbable, the chances are it will. Since no complete investigation of multiple INS failures was conducted it is possible only to speculate about such theories. Let us suggest, for argument's sake, that the number one INS begins to go wrong immediately after take-off. Since Captain Chun was the operating pilot this INS would be coupled to the autopilot to navigate the aircraft. Let us also suggest that before Nabie, where the first check of INS performance would be conducted independent of ground radio aids, the number two INS fails completely. Such incidents have been known to occur. A switch on the INS mode selector is simply set to 'attitude' to supply a stable reference for the co-pilot's instruments. The flight is free to continue in this condition although no navigational data would be available from the number two INS. INS position mixing is now lost and the Captain is in the position of the man with two watches; if they are at different times which one is correct? If, in the meantime, number three INS becomes faulty, then with two malfunctioning units and no means of cross-checking it would not be long before the pilot would unwittingly fly off track. It must be stressed that this is only speculation, although it has been shown repeatedly within this book that accidents are, more often than not, the result of such an unfortunate sequence of events. On the other hand, INS has proved itself over the years to be a remarkably reliable piece of equipment and, although an insidious sequence of errors is possible, it also does not seem probable.

So what happened to KE007? The cockpit voice recorder and flight data recorder boxes were not recovered from the depths, or so it was reported, so it will never be known. In the end, as with the Trident crash at Staines, none of the theories seems likely and it is possible only to speculate. Fortunately, some good has emerged from the incident. Communications between military and civilian controllers have been improved and plans are proceeding to link by telephone the air traffic control centres at Anchorage, Khabarovsk and Tokyo. Whether the international situation will improve as well is another matter.

Above:
Sukhoi Su-15.

Below:
KAL007 memorial in South Korea.   *Marianne Miehe*

The 747 Disasters   1985

Air-India's 747 *Kanishka*, named after Emperor Kanishka who ruled an Indian State in the second century, cruised over the mid-Atlantic at 31,000ft as it flew towards London, Heathrow. The aircraft, registration VT-EFO, was operating Flight 182 on the eastbound journey of a round trip between India and Canada. Flight 181 from India had transitted Frankfurt when travelling westbound to Toronto, then had doubled back to Montreal. There the flight number had changed to AI182 for the return trip to London, New Delhi and Bombay. On Flight 181's Toronto to Montreal sector, therefore, some passengers were inbound to Montreal from India while others were outbound from Toronto on the way to India.

Air-India's Canadian flight was a weekly service, and the present crew of 22, under the command of Captain Hanse Singh Narendra, had spent a pleasant six-day stop-over in Toronto before boarding the aircraft. The co-pilot was Satninder Singh Bhinder, also a captain but on this flight sitting in the right-hand seat, and the flight engineer was Dara Dumasia, about to retire and completing his last trip. The 19 flight attendants of Captain Narendra's crew were under the charge of Sampath Lazer.

A large expatriate Indian community had settled in Canada and Flight 182 was over three-quarters full with 307 passengers who were mostly returning on a visit to India from their adopted country. The large crew on board brought the total on the aircraft to 329 people. The time was now 06.00hrs GMT on Sunday 23 June 1985, about 2½ hours from landing, and the aircraft was estimated to arrive at Heathrow at 08.33hrs. The Air-India flight was running about 1¾ hours late because of the time taken in Toronto to fit a 'fifth pod', or spare engine. On 8 June an Air-India aircraft had an engine failure after take-off and had landed back at Toronto where an Air Canada engine had been borrowed for the homeward journey. The engine had been returned one week later and *Kanishka* now flew back with the broken engine for repair in India. The carriage of a 'spare' engine, which is fitted below the left wing between the inboard engine and the fuselage, is more commonplace than most passengers realise, and is a convenient way of transporting such a bulky item. The engine is shrouded with fairings to reduce drag and slight trim adjustments are made to maintain balanced flight. A maximum indicated airspeed is imposed with the carriage of a fifth engine, but otherwise flying characteristics are normal. Captain Narendra had requested a reduced cruising Mach number of 0.81 on the North Atlantic track for the purposes of Flight 182, instead of the normal 0.84 Mach cruise, to comply with the restricted speed.

Above:
Air-India 747. *Via John Stroud*

Below:
Rescue workers search the Japan Air Lines' wreckage for survivors. *Associated Press*

The flight time from Montreal to London was 6¼ hours, and at the beginning of the journey the passengers were served drinks and a meal. For those who could remain awake a Hindi movie was showing, but most dozed quietly in the smooth flying conditions. Outside the air temperature measured −47°C. In the cockpit, the flight crew looked out over a clear sky, and as the sun peeked its head above the eastern horizon, some low lying cloud could be seen far below. Apart from the delay over the fitting of the 'fifth pod' all was normal and routine.

Six thousand miles away on the other side of the world, and in another time zone, ground staff at Tokyo's Narita Airport unloaded baggage containers from Canadian Pacific Air Line's Flight 003 which had recently arrived from Vancouver. With good winds on the 10-hour trip the 747 had made up some time and had landed 10min ahead of schedule at 14.15hrs local Japanese time (GMT+9). Trucks ferried the containers to the ground floor of the terminal building and luggage handlers removed the bags for passenger collection. The scene was typical of a busy and noisy international airport, but it was not to remain so for long. At 15.20hrs local time (06.20hrs GMT), as bags were being unloaded from a container, one piece of luggage exploded causing a blast which shook the entire airport. A hole was blown in the concrete floor, and the unloading area was extensively damaged. Two Japanese airport staff were killed and another four seriously injured. CP Air's 747, Flight 003 from Vancouver, had arrived with a total of 390 people on board, and had the aircraft been just half an hour late there would have been a terrible disaster. There was no doubt that the force of the blast was sufficient to cause the destruction of something even as large as a 747.

Over the Atlantic, Air-India Flight 182 continued on its way to London, blissfully unaware of the events unfolding. All airlines, of course, are subject to the threat of sabotage, although some more than others, but most have implemented careful checks to safeguard against such a happening. Air-India was no exception. Internal strife in the northern State of Punjab, brought about by extremist demands for a separate nation of Khalistan, had created civil unrest in India. The trouble came to a head in June 1984 in Amritsar with the Indian Army's storming of the Golden Temple, the Sikhs' holiest of shrines. The result was a bloodbath. Sikhs throughout the world were horrified by such an act. In retaliation, the Indian prime minister, Indira Gandhi, was assassinated by her own Sikh bodyguards. Her death stunned the world and created a Hindu backlash, which resulted in Sikhs being massacred in the streets of New Delhi.

Mrs Gandhi's successor as prime minister, her son Rajiv Gandhi, was no less at risk. On a planned visit to Washington earlier in the summer of 1985 the FBI had foiled an assassination plot by Sikh terrorists. Two suspects wanted for questioning, Lal Singh and Ammand Singh, escaped capture. The proliferation of the name Singh, meaning 'the Lion', may have compounded the problems of detection, for all male Sikhs carry the name.

As a result of the struggle in India, the Indian Government was not short of enemies and Sikh extremists posed a specific threat. Air-India, as a long arm of the nation, was particularly vulnerable to terrorist attacks and was fully aware of the risks. Canada and the UK both contained the largest concentration of Sikhs outside India, and were subject to special precautions. Air-India had implemented a security system which appeared effective, and passengers boarding Flights 181/2 in both Toronto and Montreal underwent strict security measures. Air-India employed the services of local security companies who, together with the airline's own agents and the Royal Canadian Mounted Police (RCMP), subjected the passengers to a double security check. Metal detectors were used to screen for weapons and all hand baggage was searched. Suitcases were individually X-rayed and where suspicious items were uncovered a portable bomb 'sniffer' could be used to detect explosives. Three bags containing doubtful packages had been left behind in Montreal: later inspection revealed them to be safe. Only an iron, a radio and a hair dryer were found. Air Canada, as Air-India's handling agent, used a recommended passenger numbering system which ensured that all who had checked in boarded the aircraft. The security system implemented by the Indian airline seemed a reasonable and adequate response to the risks.

At 07.05hrs GMT, Air-India Flight 182 passed track position 50°N 15°W, and relayed the information to Shannon. The aircraft was just within VHF radio range and the position report was transmitted to control on 135.6MHz, a frequency which had been previously assigned on HF radio. The frequency, in fact, had been incorrectly allocated and AI182 was now instructed to call Shannon on 131.15MHz. On frequency changeover a stream of calls could be heard but eventually at 07.08:28hrs, Captain Bhinder, acting as co-pilot, established contact.

Capt Bhinder R/T: 'Air-India 182, good morning.'

Shannon Control R/T: 'Air-India 182, good morning. Squawk two zero zero five, and go ahead please.'

Capt Bhinder R/T: 'Three zero zero five squawking, and Air-India is five one north one five west at zero seven zero five, level three one zero, estimate FIR (Flight Information Region) five one north zero eight west at zero seven three five, and Bunty next.'

Shannon Control R/T: 'Air-India, Shannon, Roger. Cleared London via five one north zero eight west, Bunty, upper blue 40 to Merley, upper red 37 to Ibsley, flight level three one zero.'

Captain Bhinder repeated the instruction then Shannon replied correcting the earlier mistake and confirming the squawk of 2005.

Capt Bhinder R/T: 'Right, Sir. Squawking two zero zero five, 182.'

The time was now 07.10hrs and, with fair westerly winds, *Kanishka* flew on at a ground speed of 519kt, heading 098° magnetic towards the next position of 51°N 08°W, which lay about 50 miles south of Cork in the Irish Republic. Flight 182's routeing then proceeded up the mouth of the Bristol Channel, on across the West Country to the VOR radio beacon at Ibsley, and from there it would continue on to London.

On the flight deck the discussion centred around the flight purser's requirement for bar seals to lock bars in keeping with customs regulations. F/E Dumasia asked Captain Bhinder to radio ahead to London operations with the request. Meanwhile, in the Shannon Air Traffic Control Centre (ATCC), controllers M. Quinn and T. Lane monitored Air-India's progress, together with other aircraft in the vicinity.

Momentarily a clicking sound of a transit button came over their headsets and, as they watched the screen, the Air-India radar return suddenly vanished. The time was 07.14:01hrs GMT. Unknown to the controllers, Flight 182 had disintegrated in mid-air. The tail section aft of the wings broke off, and as the aircraft plummeted towards the ocean the wings and engines detached and fell in a shower of twisted metal into the sea. In a moment *Kanishka* was gone. There was no warning and no 'mayday' call: Flight 182 simply disappeared. With contact lost the controllers, alarmed by the circumstances, requested other flights to call Air-India, but to no avail. By 07.30hrs it became obvious that the problem was serious and an emergency was declared. The emergency services were mobilised and shipping in the area of 51°N 15°W was alerted. The Irish Navy vessel, *Le Aisling*, with cargo ships in the region, among them the *Laurentian Forest*, *Ali Baba*, *Kongstein* and *West Atlantic*, converged on the location of the crash. By 09.13hrs a radio report from the *Laurentian Forest* confirmed the worst fears as wreckage and bodies were found floating on the surface. There were no survivors; all 329 people aboard had perished. The accident proved to be the worst aviation disaster over sea, and at the time the third worst disaster in aviation history. Who was to know that before the year was out the disaster would fall into fourth place.

AIR INDIA
Crash Location

☆

CORK

Above:
Air-India crash location.

An accident co-ordination centre was set up in Cork and floating wreckage and bodies recovered from the sea were taken to the Irish port. In the days that followed the accident, about 5% of the aircraft's total structure was retrieved from the sea's surface and 131 victims of the crash were brought ashore. A team of pathologists was organised to perform autopsies and arrangements were made to fly in relatives to identify the next of kin.

The vessel *Gardline Locator* from the UK, with sophisticated sonar equipment aboard, and the French cable laying vessel the *Leon Thevenin*, with its robot mini-sub *Scarab*, were dispatched to locate the flight data recorder (FDR) and cockpit voice recorder (CVR) boxes. The batteries of the acoustic beacons attached to the recorders would survive for a maximum of only 30 days. The boxes would be difficult to find and it was imperative the search was commenced quickly. By 4 July, the *Gardline Locator*'s equipment had detected signals on the sea bed and on 9 July the CVR was pin-pointed and raised to the surface by *Scarab*. The next day the FDR was located and recovered. It was a remarkable achievement. The two boxes were brought ashore and dispatched to India for analysis.

The remaining wreckage of Flight 182 lay on the sea bed at a depth of 6,700ft and its retrieval would be difficult if not impossible. In preparation for a recovery attempt the Canadian Coast Guard vessel *John Cabot* began combing the area, taking video film of the debris on the bottom and shooting thousands of still photographs. Over the month of July, fortunately in unusually calm weather, the painstaking process of mapping the wreckage distribution was begun. It would be many weeks before it was completed.

On 16 July, the CVR and the FDR boxes were opened in Bombay and their contents analysed in the presence of international safety experts. The results were startling. At precisely 07.13:01hrs, the exact moment of the break-up, both recordings had stopped abruptly. Flight 182's electrical power supplying vital components had been completely and instantly severed. The electrics bay must have been totally destroyed. This sudden loss of electrical power was in keeping with analysis of the Shannon ATCC tape and with the abrupt disappearance of the radar target. Whatever had happened at 31,000ft out over the Atlantic was sudden and catastrophic indeed.

Meanwhile, in Canada and Japan, a full-scale investigation of the Air-India crash and the blast at Narita was being instigated by RCMP and Japanese police. At first glance there appeared little to connect the two incidents, although Canada obviously seemed to be the linking factor. If *Kanishka* had been destroyed by a bomb, the answer could lie in Toronto or Montreal, the departure points of Air-India's 181/2, or in Vancouver, the departure city of CP Air's 003. As the weeks of July passed the police evidence began to mount. An examination of passenger lists and computer records indicated that a traveller by the name of L. Singh had checked in at Vancouver but had failed to board CP Air's Flight 003 to Tokyo's Narita Airport. L. Singh was also booked on Air-India Flight 301 from Narita to Bangkok. Another passenger, M. Singh, had also checked in at Vancouver for CP Air's Flight 060 to Toronto, and he had failed to turn up as well. In both instances their bags had been loaded. M. Singh had not been confirmed on Flight 182 because of overbooking at the time of reserving his seat, but he was wait-listed for the trip. It was not permitted to check straight through, or interline, a piece of luggage onto a flight for which a passenger was only wait-listed, so what had become of M. Singh and his bag? And where had L. Singh gone? The Canadian investigation also began to unravel a confusing sequence of bookings which had been made in the name of various Singhs, including one A. Singh, in the days leading up to the tragedy. The situation was proving to be suspicious, to say the least. The vast majority of the Sikhs in Vancouver were hard working, law abiding citizens, but the plot to assassinate Rajiv Gandhi in the US indicated that extremist elements did exist in such communities. In fact, two of the names used in booking flights matched the names of the two Sikhs wanted by the FBI. It was doubtful if those implicated in the scheme to kill Gandhi were connected with events in Vancouver, but the names in which flights were booked seemed to have been deliberately chosen to advertise the fact that a Sikh terrorist group was involved. If Flight 182 had been downed by a bomb, the motives for sabotage were becoming clear. Yet one strange fact confused the inquiry: no Sikh extremist organisation claimed responsibility. On that front there was total silence.

Other causes of the demise of Flight 182 had also to be considered and examined. If results were not forthcoming from the various investigations the answer could still lie at the bottom of the Atlantic Ocean. Problems with the 'fifth pod' were dismissed with the preliminary inquiry, but one other obvious source of the tragedy, almost too shocking to contemplate with over 600 747s flying the skies of the globe daily, could be some kind of catastrophic structural failure. If such an event had occurred, other 747s throughout the world could be at serious risk.

Back in Japan, on Monday 12 August, this time at Tokyo's Haneda Airport, crowds thronged in the terminal building on a hot and sticky evening. The three-day observance of Bon, an event almost as festive as New Year's celebrations, was to begin the next day, and many travellers were flying to reunions with families for the occasion. Japan Air Lines' (JAL) Flight 123, scheduled to depart at 18.00hrs local, was fully booked for the one-hour trip to Osaka. The aircraft, a 747SR (short range), registration JA8119, was specifically constructed by Boeing for the Japanese internal market, and was built to withstand the rigours of frequent take-offs and landings.

As JA8119 departed Gate 18, 509 of the 528 seats were occupied, mostly by Japanese, but amongst the number were 21 foreigners. The 15 crew, under the command of Captain Masami Takahama, brought the total number of people on board to 524. Captain Takahama was a training captain of three years' standing and on this flight his co-pilot was under training for promotion to command. The first officer, therefore, occupied the left-hand seat in the position of acting captain, while Takahama supervised from the right. The flight plan route for JA8119 from Haneda was via Mihara, Hakone, Seaperch, airway W27 to Kushimoto, airway V55 to Shinota, then on to Osaka. The planned flight level was 240 and the estimated flight time was 54min.

Flight 123 lifted from runway 15 left at 18.12hrs local time, and departure and climb to 24,000ft were normal. The big jet flew southwest over Sagami Bay, then turned almost due west over Oshima Island as it proceeded on a direct routeing to Seaperch. JA8119 levelled off at cruise height at 18.24hrs. Almost immediately the flight began to go seriously wrong. As Flight 123 approached the eastern coast of Izu Peninsula a loud bang was heard from the rear of the aircraft. The 'beep, beep' of the cabin pressure warning horn began to sound, then stopped.

18.24:34 Flight Engineer (F/E): 'No.'

Capt Takahama: 'Did you find anything? Check the gear.'

At the rear of the cabin it was obvious a serious event had occurred and Yumi Ochiai, a 26-year-old off-duty JAL stewardess, witnessed the scene.

'There was a sudden bang. It was overhead in the rear. My ears hurt. Immediately the inside of the cabin became white.'

A section of the aft fuselage had fractured causing the cabin to depressurise and to mist the atmosphere.

'No sound of an explosion was heard', she continued. 'The ceiling above the rear lavatory came off.'

Immediately oxygen masks dropped from the roof and an announcement automatically operated to instruct passengers in their use. It was a most alarming experience for those on board. On the flight deck it was equally disturbing for the crew, for there was no doubt that a dire emergency was developing. The captain called for the emergency squawk of 7700 to be selected on the transponder.

18.24:46 F/E: 'Hydraulic pressure down . . . amber light on. . . .'

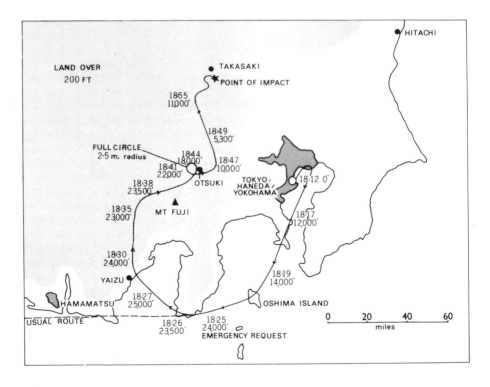

Above:
JAL123 flight route.

222

The cabin pressure warning horn began to 'beep' again. Capt Takahama called for a right turn but the controls seemed to have little effect. 'Right turn', he asked again.

The first officer exercised the control column and rudder, but to no effect. 'I did', he replied.

Whatever damage had occurred at the tail, it appeared to have fractured all four hydraulic lines, for none of the flying controls operated.

18.25:20 Capt Takahama R/T: 'Ah Tokyo, JAL 123. Request for immed... e... trouble. Request return to Haneda. Descend and maintain two two zero. Over.'

Tokyo Control R/T: 'Roger, approved as you requested.'

Capt Takahama R/T: 'Radar vector to Oshima, please.'

The flying controls no longer had any effect and the aircraft began to oscillate out of control, pitching, rolling and yawing all at the same time. The flight condition is known as a 'Dutch roll', and is a most uncomfortable manouvre for all on board. Flight 123 was unable to maintain a constant altitude and began to descend, then climb.

18.26:25 Capt Takahama: 'Hydro . . . all no good.'

18.28:30 Tokyo Control R/T: 'Fly heading zero nine zero, radar vector Oshima.'

Capt Takahama R/T: 'But, now uncontrol.'

18.28:39 Tokyo Control R/T: Uncontrol. Roger, understood.'

The aircraft flew across Suruga Bay, and as it did so door five right warning light illuminated indicating that it was unlocked. Flight 123 then turned northwards and proceeded inland towards Mount Fuji. The interphone chimed in the cockpit and a flight attendant from the back of the aircraft reported the damage to the flight engineer.

F/E Interphone: 'Yes, what is it? More rearward? Er, what is broken? Where? Ah, ah, ah. It is around the baggage hold, is it? It is in the rear, the rearmost? Yes, I understand. Well the baggage hold . . . the baggage area has collapsed. Is it door five right window? I understand.'

The flight engineer replaced the interphone handset and spoke to his captain.

18.33:33 F/E: 'Captain, door five right window. Emergency descent may be advisable. Shall we use our oxygen masks, too?'

Capt Takahama: 'Yes, that's better.'

Tokyo Control R/T: 'You are now 72 miles from Nagoya; can you land at Nagoya?'

Capt Takahama: 'Request return to Haneda.'

With all flying controls and hydraulic systems lost the aircraft was in a desperate situation and as the pilots fought to gain control the captain called a continuous series of instructions to co-ordinate their efforts. The engines still operated normally and engine power was their only means of guidance: a far from simple task. The aircraft was also still 'Dutch rolling' about the sky. Juggling with the power the pilots managed to turn onto an easterly heading and, as they passed to the north of Mount Fuji, gingerly attempted to commence descent. A problem arose in controlling the fast speed and the captain suggested lowering the gear. Since all hydraulic power was gone, the alternative system was used and up-locks were electrically released in order to allow the gear simply to fall into place by gravity.

The lowered gear seemed to dampen the motion a bit, and Flight 123 settled in descent. At about 20,000ft, however, Capt Takahama momentarily lost complete control and the aircraft banked sharply and turned a full circle in three minutes over Otsuki City before rolling out again on an easterly course. Passing 10,000ft JA8119, then turned north again towards the mountainous terrain.

18.47:15 Tokyo Control R/T: 'Can you control now.' (Japanese)

Capt Takahama R/T: 'Uncontrollable.' (Japanese)

The crew now became alarmed with approaching high ground they could see.

18.47:28 Capt Takahama: 'Hey, there's a mountain.'

18.47:45 Capt Takahama: 'Turn right. Up. We'll crash into a mountain.'

As full power was applied to arrest the rate of descent the aircraft began to gyrate wildly, pitching up and down, with the speed increasing and decreasing rapidly. Desperate attempts were made to get some response from the controls but the effort was in vain. Suddenly the speed began to drop rapidly.

18.48:49 F/E: 'Shall I rev it up?'

Capt Takahama: 'Rev up, rev up.'

'Oh, no.'

'Stall.'

The speed dropped to 108kt.

Capt Takahama: 'Maximum power.'

F/E: 'We are gaining speed.'

Capt Takahama: 'Keep trying.'

The speed increased then dropped back, and fluctuated around 200kt.

18.50:55 Capt Takahama: 'The speed is 220kt.'

In spite of the gallant effort to control pitch, the aircraft continued to descend and the flaps were lowered on the alternative electrical system to 5° to help aid recovery. The process took almost four minutes, but had the desired effect and with increased power the aircraft managed to start climbing. At about 18.53hrs the altitude peaked at 11,000ft and Flight 123 began to descend again. Captain Takahama changed frequency to Tokyo Approach Control and verified that the aircraft was out of control. He then requested his position.

Tokyo Approach R/T: 'JAL 123, your position five ah, five ah, 45 miles northwest of Haneda.'

Capt Takahama R/T: 'Northwest of Haneda. Eh, how, how many miles?' (Japanese)

Tokyo Approach R/T: 'Yes, that's right. According to our radar it is 55 miles northwest, ah 25 miles west of Kumagaya. Roger, I will talk in Japanese. We are ready for your approach any time. Also, Yokota landing is available. Let us know your intentions.'

The rate of descent now increased to 1,350ft/min with the mountains dangerously close. Using flaps and power the crew tried desperately to control the stricken machine. It was amazing the aircraft had stayed in the air for so long but it was doubtful if it could do so for much longer.

18.55:04 Capt Takahama: 'Flap set?'

F/O: 'Yes, flap ten.'

18.55:15 Capt Takahama: 'Nose up.'

'Nose up.'

'Nose up.'

18.55:42 Capt Takahama: 'Hey, hold the flap . . . ah, don't lower so much flap. Flap up, flap up, flap up.'

18.55:55 Capt Takahama: 'Power, power . . . flaps?'

F/O: 'It is up.'

18.56:05 Capt Takahama: 'Nose up.'

'Nose up.'

'*Power.*'

At 18.56:15 the aircraft took a final plunge earthward and the ground proximity warning system (GPWS) sounded, alerting the crew to the approach of high ground.

GPWS: 'Pull up, pull up, pu… pu u… .'

Nothing could be done. The crew were helpless in the situation and at 18.56:20 the flight of JA8119 ended. The aircraft crashed at a height of 4,780ft into the side of Mount Osutaka at about 70 miles northwest of Tokyo in the Gumma prefecture. JA8119 almost totally disintegrated on impact and burst into flames as fuel tanks ignited. It was a disaster of monumental proportions.

Flight 123 had crashed in a remote mountain forest in an area where not even a helicopter could land safely. In the gathering darkness, units of the Japan Air Self-Defence Force made passes over the site but no sign of life could be seen in the wreckage. A team of local firemen set out during the rainy night to negotiate the difficult climb to the crash scene, while a base was set up in a mountain village to co-ordinate efforts. It wasn't until nearly 09.00hrs local time the next day, over 14 hours after the crash, that the firemen reached the area, and were joined by paratroopers sliding down ropes from helicopters hovering overhead. The aircraft had broken apart on a narrow ridge and sections had catapulted into ravines on either side. One fireman searching in a gully detected movement in the debris and to his amazement found Yumi Ochiai, the off-duty JAL stewardess, trapped but alive in the wreckage. Three others also escaped death: 12-year-old Keiko Kawakami, Hiroko Yoshizaki and her eight-year-old daughter, Mikiko. It was quite astounding that anyone had survived. These four, seated together in the centre of row 56, were the lucky ones, for of the 524 on board, 520 had perished; the worst single accident in aviation history and second only to the carnage at Tenerife in 1977. As the grim task of clearing the bodies from the wreckage in the forest began, notes of farewell, testifying to the horror endured by the passengers during the last 30 minutes of flight, were found amongst the dead.

News of the disaster shocked the Japanese nation. The JAL president, Yasumoto Takagi, resigned in a gesture of corporate 'guilt' and shouldered full responsibility for the crash. A JAL maintenance manager at Haneda, shamed by the high death toll, took his own life in 'apology' for the disaster.

The summer of 1985 proved to be one of the blackest times in aviation history. In only two accidents, two months apart, two 747s apparently suffering from some kind of structural damage had crashed killing 849 people. It was as if history was repeating the tragedies of the Comet crashes on a grand scale. Not surprisingly the aviation world was deeply alarmed by events. If the 747 suddenly and inexplicably was proving to be structurally unsafe, answers had to be found, and found quickly.

On the day following Flight 123's demise, a first clue as to the cause of the accident was revealed when parts of JA8119's tail fin and lower rudder were found floating in Sagami Bay and were recovered by a Japanese destroyer. Panelling from the rear fuselage and ducting from the auxiliary power unit were also retrieved. It was discovered that almost half the tail fin had detached in flight and it was amazing that the aircraft had stayed in the air for so long. The initial suspicion that door five right was the source of the trouble was quickly discounted when it was found intact at the crash site, and the cause of the damage remained a mystery. Examination of safety records, however, revealed that in June 1978 a tail scrape had occurred when JA8119 had landed at a high nose-up attitude in Osaka. The rear pressure bulkhead, an aluminium-alloy umbrella shaped dome which plugs in the end of the pressurised cabin, was damaged and was repaired in Japan by Boeing engineers. Could there be a connection? During the weeks of August, as accident investigators busied themselves with the causes of the two 747 crashes, and experts in Japan and India analysed the data, the aviation world held its breath. Checks were conducted on all 747 tail fins and rear pressure bulkheads. Photographs of Air-India Flight 182's tail section, lying at the bottom of the Atlantic, were scrutinised for signs of damage in an attempt to establish some link between the crashes, but any fear of a connection was dispelled when it was found intact.

In the Atlantic work, continued on through August and September to locate and map the distribution of Air-India's wreckage on the sea bed. The Atlantic floor at the site lay at a shallow angle, but there were no obstructions. Torn sections rested on the soft silt as if on a beach, within a radius of five to six miles. The forward aircraft section lay inverted at the beginning of the wreckage trail and was badly damaged, while the fuselage pieces aft of the wings were found strewn over a five-mile path. The forward cargo door was located detached, bent and damaged, the wings were discovered together with one lying flat and the other near vertical, while all engines had completely separated. As a detailed map of the wreckage was completed, experts earmarked what were considered important sections for the attempted salvage operation expected to begin in October. Pieces recovered from the deep, it was hoped, would shed new light on the subject for as yet no concrete proof of the cause of the accident had emerged. The distribution of debris on the sea bed seemed to indicate that if the break-up had been the result of a blast, the bomb was likely to have been placed in the rear cargo hold. Identification of passengers and pathological reports in Cork, however, indicated that the bodies retrieved from the sea had been thrown from the rear section and none had shown signs of the effect of an explosive device. The sudden loss of electrical power and photographs of the front fuselage section and the detached forward cargo door damage indicated an explosion in the forward cargo hold. The other cargo doors were noticed to be intact and still attached to the fuselage. Scientists studying the CVR tapes were also in disagreement regarding the final milliseconds of the recording. Some stated the trace indicated an explosion at the forward end of the aircraft while others declared the findings inconclusive on that count. The recording, the dissenters claimed, was more consistent with an explosive decompression, such as a cargo door blowing out, than an explosive device.

226

Meanwhile, in Canada, the work of the RCMP was beginning to bear fruit. The bewildering series of bookings in the name of Singh in the few days leading to the crash of 23 June had been unravelled and circumstantial and criminal evidence strongly supported the bomb theory. On 16 June an A. Singh originally made a reservation in Vancouver for a flight on 22 June to Tokyo on CP Air 003, to connect with Air-India 301 to Bangkok at Narita. The ticket was never purchased. On 19 June, a booking was made for 22 June on CP Air 060 to Toronto to connect with Air-India 181/2 to New Delhi in the name of Jaswand Singh. At the same time a return ticket on the same date was made on CP Air 003 to Tokyo connecting with Air-India 301 to Bangkok in the name of Mohinderbel Singh. On 20 June, a man of 'East Indian' appearance paid over $3,000 in cash for the tickets at CP Air's Vancouver office after changing Mohinderbel's journey to the name of L. Singh and Jaswand's journey to the name of M. Singh. He also changed the Vancouver-Tokyo-Bangkok return to a single ticket. Since Air-India 181/2 was overbooked, it was only possible to wait-list M. Singh on the flight out of Toronto to Montreal and Delhi.

In Vancouver, on the morning of 22 June, M. Singh checked-in for CP060 to Toronto. An argument ensued at the check-in counter as Singh demanded his bag be checked through to Air-India 181/2 although his booking couldn't be verified. The agent refused to accept the bag, but M. Singh insisted that he had just phoned Air-India and that he was confirmed on the flight. Unable to check the details on her computer console, and with a large line forming at her wicket, the agent relented and interlined his suitcase to Flight 181/2. At 16.18hrs GMT, CP060 departed for Toronto without M. Singh. Later L. Singh checked in for CP003 and his suitcase was interlined to Air-India 301. CP003 departed for Tokyo without L. Singh at 20.37hrs GMT. Five minutes later CP060 arrived in Toronto and the bags were transferred to AI181/2. As the luggage was being X-rayed before loading the machine broke down and the staff resorted to the use of the PD-4 hand-held explosives 'sniffer' to check for bombs. The security staff present claimed not to have received proper instruction in the use of this equipment, but with the X-ray machine broken there was little choice. An Air-India staff member demonstrated its used by holding a match to the sniffer and it gave off a loud whistling noise. On checking the bags for AI181/2 the PD-4 did 'beep' softly when passed over the lock of one suitcase, but nothing like the sound made when demonstrated. No attempt was made to depress the bag to expel air for the equipment to 'sniff'. This bag also seemed to match one checked-in by M. Singh in Vancouver. The Toronto-Delhi bags were loaded in pallets spread through the holds and also loosely in the bulk cargo hold at the tail, but of the baggage containers two were known to have been placed in the forward cargo hold aft of the electrics bay. The security system for passenger check-in implemented by Air-India at Toronto and conducted by Air Canada staff did not include interlining passengers and M. Singh's absence was not detected.

AI181 departed Toronto at 00.15hrs GMT on 23 June, arrived Montreal, Mirabel, at 01.00hrs and, after a transit of over an hour, the aircraft departed for London as Flight 182 at 02.18hrs. As AI182 commenced its Atlantic crossing, CP003 arrived at Tokyo at 05.41hrs. At 06.19hrs the baggage container exploded on the ground floor of the terminal building. Less than one hour later at 07.14hrs

GMT, VT-EFO disintegrated and disappeared from radar screens. The bomb in Tokyo had been concealed in the frame of an AM/FM stereo tuner and Japanese forensic experts had not only managed to identify the manufacturer as Sanyo, but had also established that the model number was FMT 611K. All 2,000 of these units, produced in 1979, had been dispatched from the factory to Vancouver. Its size was bigger than more recent equipment and was just large enough to contain the explosives and timer. After much foot-slogging the RCMP discovered the store which had sold the out-of-date model to two Sikhs only three weeks earlier. Unfortunately, only one such sale was uncovered, but the other tuner may very well have been in one of the perpetrators' homes and may have sparked the idea to use one of that size.

An examination of airlines' timetables indicated that the Sikh terrorists had probably intended the bombs to detonate simultaneously aboard the Air-India flights while the aircraft were on the ground at Tokyo and London. Since timetables state arrivals and departures in local time, the Tokyo bomb could have been mis-set by one hour. Canada and the UK both advance their clocks by one hour for daylight saving in the summer while Japan does not, and a one-hour discrepancy could have resulted when synchronising the timing devices. The date of the attacks was also close to the first anniversary of the storming of the Golden

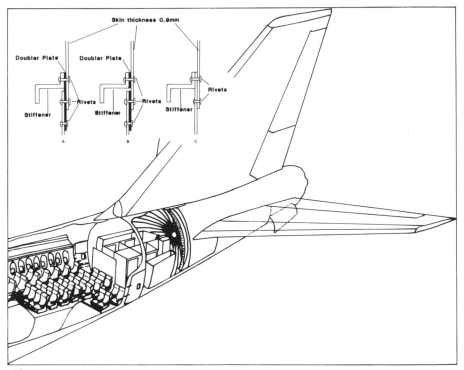

Above:
The 747 layout and pressure bulkhead details showing, (a) correctly mended section;
(b) incorrectly mended section (note gap of doubler plate between top two rivets) and
(c) the original bulkhead.

228

Temple in June 1984, and gave a clear motive for the outrages. Had the Tokyo bomb been timed to detonate one hour later and AI182 not been delayed 1¾ hours by the fitting of the spare engine, the terrorists may very well have achieved their goal. It would have been a dire and relatively sophisticated warning to the Indian Government that Sikh extremist organisations could strike at will throughout the world wherever and whenever they chose. The plot, however, appeared to have been too ambitious and the plans went badly wrong. It was perhaps now not surprising that no group was prepared to claim responsibility for the indiscriminate slaughter of 329 innocent victims aboard the flight, and the killing of the two ground staff at Narita.

In Japan, the investigation into the JAL crash continued and reached a satisfactory conclusion by the end of September. Boeing, in an admirably open gesture, admitted that the rear pressure bulkhead of JA8119, which had been damaged in the tail scrape incident at Osaka in 1978, had been incorrectly repaired by their engineers. On the departure out of Haneda in August the bulkhead had ruptured and air spilling from the breach had pressurised the tail fin resulting in disintegration of the fin and lower rudder. In the process the four sets of hydraulic lines running through the bulkhead to the empennage controls had severed. All hydraulic systems were rendered useless as fluid spilled from the lines and within 30 seconds the flying controls were lost. A joint fund was set up by Boeing and JAL on a 50/50 basis to compensate victims of the disaster, although no recognition of liability was accepted by either company. The investigation's conclusion broke any suspicion of a link between the two 747 disasters and upheld the integrity of the 747 fleet as a whole. Meanwhile, JAL continued with thorough structural checks on all its 747s. The process would take some time but if any other faults lay hidden in the structure they would be found. The following year JAL were to be rewarded for their efforts!

As the circumstantial and criminal evidence in favour of the Air-India bomb theory mounted, work began in October to retrieve wreckage from the Atlantic sea bed. In an amazing salvage operation lasting over two months, the *John Cabot* and the mini-sub *Scarab*, assisted by the vessel *Kreuztrum*, a Canadian offshore supply ship, plucked over 20 significant pieces, including some large sections, from the depths. The badly damaged forward cargo door was retrieved, but as it was being lifted clear a gust of wind blew it from the line and it was never recovered. Puncture holes in pieces of wreckage seemed to verify the bomb theory, but on closer scrutiny scientists were unable to agree on the findings. The damage caused by an explosive device resulting in structural failure and subsequent rapid decompression is similar to structural failure leading to an explosive decompression. No chemical evidence of an explosion or bomb fragments was found. Further analysis of debris, however, revealed scorch marks underneath seat cushions, seat legs buckled, and a cabin floor section in a cupboard deformed in a dome-shaped fashion. A Canadian Air Safety Board (CASB) report released at the end of January 1986 supported the bomb theory, although admitting it could not be positive. 'There is considerable circumstantial and other evidence', the report stated, 'to indicate that the initial event was an explosion occurring in the forward cargo compartment. This evidence is not conclusive. However, the evidence does not support any other conclusion.' The official Indian inquiry into the crash opened in Delhi in mid-November 1985 under Judge Bhupinder Kirpal, and interim reports issued over the period of its sitting also supported the bomb theory. Before the concluding report could be published in 1986, however, the integrity of the 747 design was again cast into doubt with further revelations in January. In Tokyo, JAL engineers, still carrying out thorough checks in the aftermath of the disaster at Mt Osutaka, discovered cracked and broken ribs in the frames of the forward sections of some 747 fuselages. The disclosure was another blow to the prestige of the 747 just as the consternation over the 747 crashes in 1985 was beginning to subside. The 747 was originally designed with the flight deck placed atop the fuselage to permit nose loading of cargo containers, and resulted in a unique pear-shaped forward cross-section. The cracks were attributed to fatigue caused by cabin pressurisation stresses on this rather singular structure. The FAA issued an emergency directive requiring a worldwide inspection of the fleet. At the time 610 747s were being operated by 69 carriers. All aircraft having completed 10-14,000 flights had to be examined within 50 landings, and those having completed 14,000 flights or more had to be checked within 25 landings. The importance of the directive was stressed by the FAA stating that 'failure of adjacent frames could lead to rapid decompression and could possibly cause the loss of an aeroplane'. Amid some alarm, more 747s with nose frame cracks were found by other airlines. Two weeks later Boeing also issued a service bulletin recommending internal visual inspection.

The revelations of forward frame cracking raised further questions as far as the Air-India crash was concerned, and prompted suggestions that VT-EFO's nose section, still lying on the Atlantic bottom, should be raised to dispel any doubts. The size of the structure, however, would make lifting very difficult. Engineers also stated that nose frame cracking leading to explosive decompression in an aircraft so young as the Air-India 747 was 'extremely improbable'. VT-EFO was only seven years old and had flown 23,634 hours in only 7,522 flights. By comparison, the short sector JAL 747s had completed more than double the number of flights. JA 8119, for example, was 11 years old and had accumulated 25,025 hours in 18,830 flights.

Detailed aircraft inspection and prompt repair work overcame the frame-cracking problem and outlined the success of the structural monitoring programme which had resulted in initial exposure of the nose frame cracks. In Canada, the RCMP continued its investigation of Sikh terrorist groups and arrests of Sikh extremist suspects were made, although no positive links with the sabotage of VT-EFO were established. No claim by any Sikh terrorist organisation was ever made. The Air-India break-up, however, was officially attributed to a bomb. The JAL crash was recognised as a maintenance error. After a year of difficulties, the integrity of the 747 design was upheld, and in spite of misgivings the 747 was verified to be a very strong aircraft. There is no doubt that it will continue to fly for many years to come.

Summary

The accidents outlined within the pages of this book tell of a terrible sequence of disasters, but lessons have been learned from these tragedies of the past and much effort has been applied to ensure a safer aviation world for all.

On 5 October 1930, the R101 set out on a dangerous journey without being properly tested and carried important passengers aboard at great risk. Contrast that with the development of Concorde, the most tried and tested aircraft ever to enter commercial service. After 10 years of operation not a single passenger has been lost. The subsequent R101 inquiry proved inadequate, but the process of accident investigation has continued to develop over the years and has now matured into a precise science. On 31 October 1950, in an accident not recounted in this book, a Vickers Viking crashed in fog at London Airport, killing 28 people. From this accident minimum visibility and cloud base limits were introduced, eventually worldwide, below which aircraft were not permitted to land. The introduction of these limits proved to be a significant advance in safety. In that same year the first flight data recorder (FDR) was developed and used in Australia. The Comet crashes of 1953 and 1954 revealed the dangers of metal fatigue and introduced improved standards of aircraft construction. Improvements to fatigue and corrosion detection procedures were introduced and today, in addition to the white-coated inspector with his torch, localised cracking can be detected by apparatus measuring the distortion of eddy current waves, aircraft sections can be X-rayed, and deep analysis of vital areas can be conducted by the use of ultra-sonic crack detection techniques. On 6 February 1958, the disaster at Munich revealed the problems of slush and eventually resulted in strict limits being imposed for take-off and landing in such conditions. The position of the house 300yd from the end of the runway at Munich, which the Ambassador aircraft struck, would now no longer be permitted and unobstructed zones at runway ends are compulsory. Although such obstructions have been removed, some hazards still exist. On 7 October 1979 a DC-8 landed well down runway 15L at Athens Airport with heavy rain falling. The runway was coated with oil and rubber and was slippery with the first rains after a long dry summer. The aircraft overran the runway and slid down a 12ft escarpment. The overrun had not been levelled off after seven years of representations from the Greek pilots' association, and 14 people were killed. If the emergency overrun had been level the aircraft would not even have been damaged. Instead of filling in the dip, the authorities sentenced the pilots to jail for manslaughter! Fortunately the pilots successfully appealed.

In 1965, in another accident not recounted in this book, a Vanguard crashed on an attempted go-around at London Airport. The aircraft carried the first British produced flight data recorder (FDR), a piece of equipment of high quality. The recorder proved its worth, producing excellent flight details and led to a most thorough investigation of the accident. The crash resulted in the first presentation of recorder details as legal evidence in a UK court. FRDs are now mandatory on all large transport aircraft. On 18 June 1972, the Trident crash at Staines resulted in the compulsory installation of cockpit voice recorders (CVR) on flight decks in the UK, a condition which has now been extended worldwide. On 3 March 1974 the DC-10 crash at Paris resulted in the strengthening of wide body floors to withstand the decompressive effect of a 20sq ft hole appearing in the fuselage. On 10 September 1976 the BEA/Inex-Adria airborne collision over Zagreb resulted in improved air traffic control (ATC) monitoring procedures and hastened the development of anti-collision equipment. Further improvements to radar surveillance are now being implemented with the introduction of mode S discretely addressed secondary surveillance radar allowing more aircraft to be handled with increased safety. Airborne anti-collision systems are also being introduced which provide full protection from aircraft equipped with a similar system and useful protection from aircraft equipped only with transponders. The system can be operated in high density environments and does not interfere with ATC procedures. To improve pilot look-out, powerful flashing white strobe lights, which can be seen even in daylight, are now standard on new aircraft. On 27 March 1977 the collision of the 747s at Tenerife resulted in the recommendation of improved radio communications, especially in the take-off phase. On certain airport charts, following a number of other ground incidents, taxiway directions have been printed to help taxying procedures, especially in reduced visibility. The installation of ground radar at airports with a fog risk was also recommended, but the equipment is expensive and its implementation is slow. On 25 May 1979 the DC-10 crash at Chicago resulted in engine failure take-off speeds being reviewed, and on 28 November of the same year the DC-10 crash on Mount Erebus resulted in the controversial Antarctic flights being withdrawn. An increased awareness of pilot visual perception was also realised. This accident inspired the development of an inertial navigation system (INS) route data checking facility which monitors systems for the detection of incorrect computer co-ordinates. A safety device which might have saved the DC-10 and other flights — a forward facing ground proximity warning system — is also being developed. The 30min cockpit voice recorder tape proved inadequate in this incident and valuable information was lost because of the restricted time. The 747 crash in Japan suffered from the same problem. In the future it may be deemed necessary to increase the time to one hour, perhaps accompanied by a video camera and recorder running simultaneously. On 1 September 1983, the downing of the 747 over Sakhalin Island resulted in improved military and civil aviation co-operation, and a ground telephone link has now been installed between US, Soviet and Japanese ATC centres. On 23 June 1985, the 747 lost off the coast of Ireland in a suspected bomb explosion resulted in increased security at high risk airports, and the 747 crash in Japan on 12 August of the same year led to improved aircraft inspection schedules. In February 1986, as a result of an intensive Japan Air Line's inspection programnme, cracking in the 747 nose section was discovered. An alert to other operators revealed the problem on more aircraft, and repairs were instigated to eliminate the potential hazard.

Improvements to safety continue relentlessly and efforts are being made in many fields to reduce risk and to increase the chance of survival in an accident. One area which has received much attention recently has been the prevention of fires and the protection of passengers from lethal fumes. An anti-misting fuel is being developed which will reduce the risk of fire on rupture of a fuel tank. In December 1984 a Boeing 720, carrying fuel with an anti-misting additive, was crashed by remote control at Edwards Air Force Base, California. The crash was the culmination of five years of planning, but was only partially successful. Work continues to perfect the additive. In aircraft fires more people are killed by inhaling smoke and fumes than by incineration. In the cabin, fire resistant and low smoke emission materials are now being used for upholstery, and smoke hoods are being placed on board for passenger protection. Emergency lighting at low level is being incorporated in the cabin to guide escapees in a smoke filled aircraft, and escape slides are being developed with aluminium coatings which can preserve the slide integrity against fire for longer periods. Improvements to cabin safety briefings have also been implemented with the introduction of excellent video presentations in some airlines. All that is required now is for the passengers to pay attention to them. In fact, passengers can play a significant role in safety. Fire on an aircraft in flight is one of the greatest hazards, and matches or lighter fuel should not be placed in baggage. Nor should any dangerous goods, such as butane cylinders or fireworks. Hand baggage should be restricted to one small piece to maintain seat areas free from obstruction and duty free spirits should be limited to one bottle per adult. There is no doubt that alcohol on board is a fire hazard and in future the carriage of duty free spirits may be banned. What are required are more airports like Singapore, Cairo and Sydney with facilities to purchase goods on arrival. Accident survival can also be improved by placing seats to face backwards, as is the case on RAF transport aircraft, but passengers are known not to like this configuration.

In the operation of aircraft, humans, both on the ground and in the air, are still proving to be the weak link in the chain, and more is being done, and needs to be done, to improve the situation. Pilots will retain their place on the flight deck for a very long time as, in spite of the advances of computers and electronics, creative thinking is still a necessary facet of the job. Further research is being conducted on the long term and the cumulative effects of fatigue, and flight time limitations for crews need to be constantly reviewed. The introduction of two-pilot crews on long haul operations poses greater problems. More is also being done to ensure the suitability of applicants, as some aviators, although technically competent, have personalities which are unsuitable to the flight deck environment. In the past psychological screening of pilots at initial selection has been minimal and on promotion from co-pilot to captain, negligible. Safety can be compromised when unsuitable individuals hold positions on a flight deck and, once established, companies find it difficult to dismiss disruptive flight crew members.

As aircraft become more automated, further research is being conducted on the interface between operator and machine. Increasingly, computers fly the aircraft while pilots monitor: a function which human beings do not perform well. There is a danger of automation complacency setting in. Automatics are extremely reliable, but they do go wrong, and their dependability can increase the danger of pilots being caught out when malfunctions occur. Greater, not less, vigilance is required. An 'electronic cocoon concept' is being developed whereby pilots play a more positive role in the operation of the aircraft and the computers do the monitoring. A most satisfactory arrangement for all!

As air traffic continues to increase, aviation research continues apace with improvements to air safety. Many projects under development today will be implemented in the next few years and will help to ensure improved standards of safety into the 1990s.

Bibliography

R101 (1930)
Sir Peter Masefield; *To Ride the Storm: the Story of the Airship R101*; Kimber, 1982.
Geoffrey Chamberlain; *Airships: Cardington*; Dalton, 1984.
'Report on the R101 Inquiry'; HMSO, 1931.
Public Records Office, Kew. References: AIR 5/902-906, 909-913 and 916-920.

Comet (1953 and 1954)
'Report on Comet Accident Investigation'; RAE, Farnborough, 1954.
UK Civil Aircraft Accident Reports: CAP 112 (Calcutta), HMSO, 1954; CAP 127 (Elba and Naples), HMSO, 1955.
Oliver Stewart; *Danger in the Air*; Routledge, 1958.
Andre Launey; *Historic Air Disasters*; Ian Allan, 1967.
Stephen Borley; *The Search for Air Safety*; Morrow, 1970.
Michael Hardwick; *The World's Greatest Air Mysteries*; Odhams, 1970.

Munich (1958)
Stanley Williamson; *The Munich Air Disaster: Captain Thain's Ordeal*; Cassiser, 1970.
Frank Taylor OBE; *The Day a Team Died*; Souvenir Press, 1983.
UK Civil Aircraft Accident Reports: CAP 153, 167, 292, 318; HMSO.
'Application of the Results of Slush Drag Tests on the Ambassador to the Accident at Munich'; RAE, Farnborough.

Trident, Staines (1972)
John Godson; *Papa India: The Trident Tragedy*; Compton Press, 1974.
UK Civil Aircraft Accident Report: 4/73; HMSO, 1973.

DC-10, Paris (1974)
John Godson; *The Rise and Fall of the DC-10*; New English Library, 1975.
The Sunday Times Insight team, Paul Eddy, Elaine Potter and Bruce Page; *Destination Disaster*; Hart Davies McGibbon, 1976.
American NTSB Report: AAR-73-2 (The Windsor Incident), 1973.
UK Civil Aircraft Accident Report: 8/76 (trans from French); HMSO, 1976.

Zagreb Collision (1976)
Richard Weston and Ronald Hurst; *Zagreb 14: Cleared to Collide*; Granada, 1982.
UK Civil Aircraft Accident Report: 5/77 (Unpublished), 9/82; HMSO, 1982.

Tenerife Collision (1977)
Spanish 'Subsecretaria de Aviacion Civil' Accident Report, 1978 (ICAO circular 153-AN/56).
American Airline Pilots' Association (ALPA) Accident Report, 1978.

DC-10, Chicago (1979)
William Norris; *The Unsafe Sky*; Arrow, 1981.
American NTSB Report: AAR-79-17, 1979.

DC-10, Mount Erebus (1979)
Kenneth Hickson; *Flight 901 to Erebus*; Whitcouls, 1980.
Gordon Vette with John Macdonald; *Impact Erebus*; Hodder & Stoughton, 1983.
Peter Mahon; *Verdict on Erebus*; Fontana, 1985.
Right Vision Ltd; *Impact Erebus Video*; Auckland, New Zealand. 1985.
New Zealand Aircraft Accident Report: AAR 79-139 (The Chippendale Report), 1980.
Royal Commission of Inquiry Report, Judge Peter Mahon, 1981.

Korean 747 Shoot-down (1983)
Richard Rohmer; *Massacre 007*; Coronet Books, 1984.
Murray Sayle; *The Sunday Times Review*; *The Sunday Times*, 20 and 27 May, 1984.
ICAO Report (Unpublished).

The 747 Accidents (1985)
Salim Jiwa; *The Death of Air-India Flight 182*; W. H. Allen, 1986.
ICAO Report: 1985-5 (Air-India).
Canadian Aviation Safety Board; January 1986 (Air-India).
Indian Aircraft Accident Inquiry Report: 1986.
ICAO Report 1985-6 (Japan Air Lines).
Japanese Civil Aircraft Accident Interim Report, August 1985.

236

Index

Wreckage litters the forest floor showing the swathe cut by the DC-10 at Paris in 1974. *Associated Press*